Classic Trains and Railroad Engineering

Volume 1

Antique Locomotive Operations Part 1

Edited by Mark Bussler

Classic Trains and Railroad Engineering: Volume 1
Antique Locomotive Operations Part 1

Restored and Edited by Mark Bussler
Cover Design by Mark Bussler

Volume 1 of 8

www.CGRpublishing.com

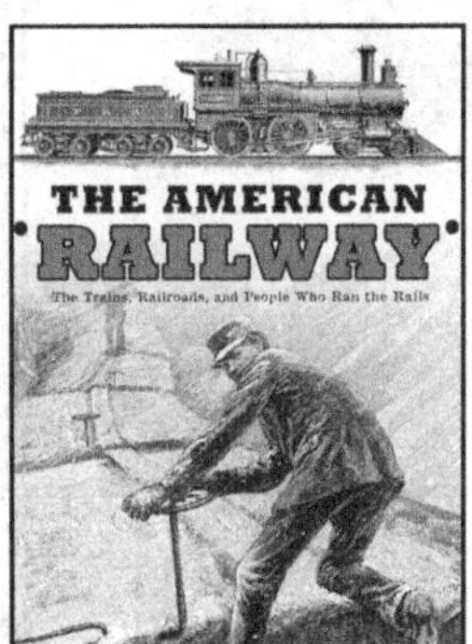

The American Railway: The Trains, Railroads, and People Who Ran the Rails

Our Iron Roads: Railroad History, Construction, and Administration

When Railroads Were New: Restored Centenary Special Edition

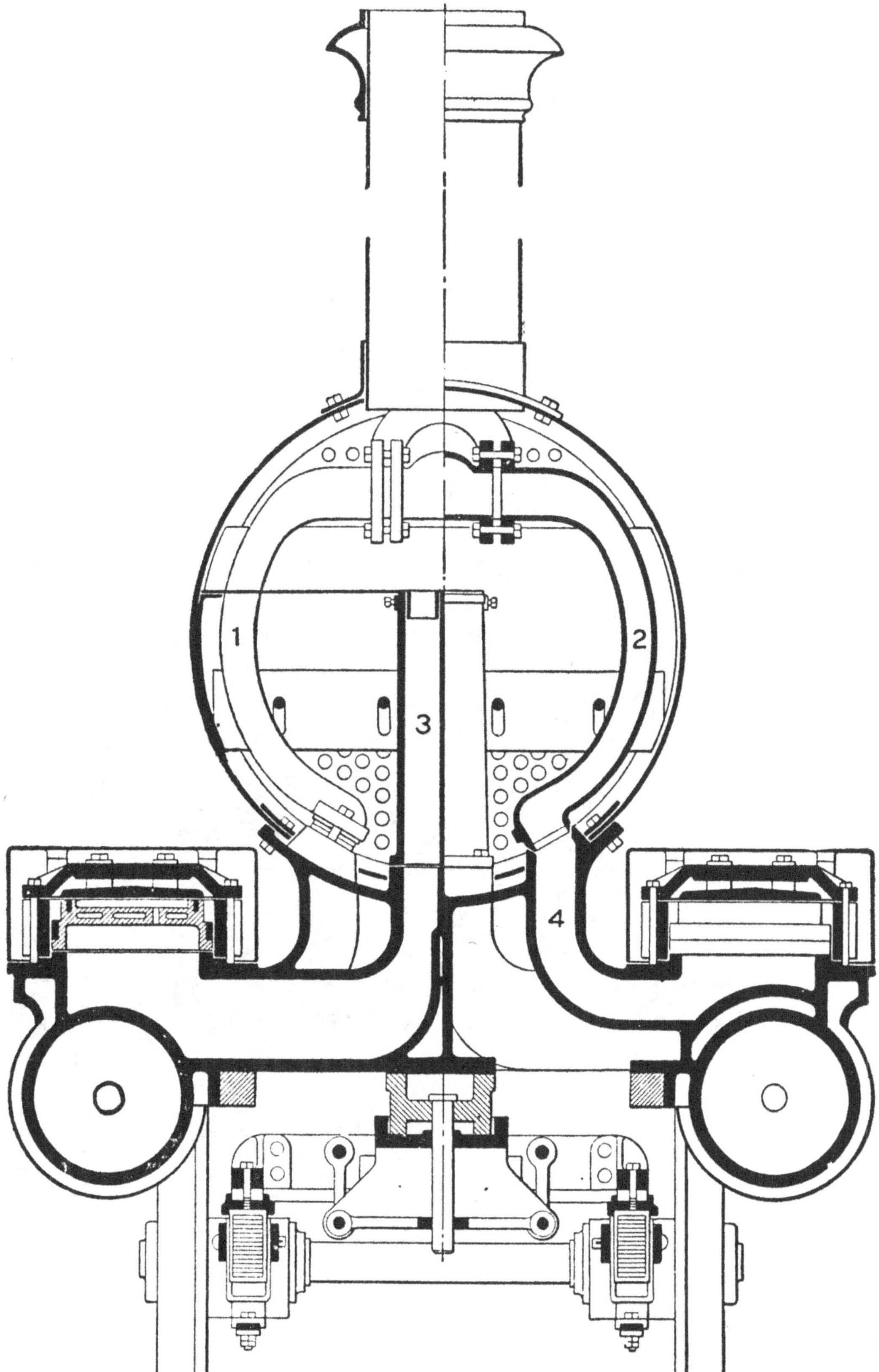

FIGURE 31

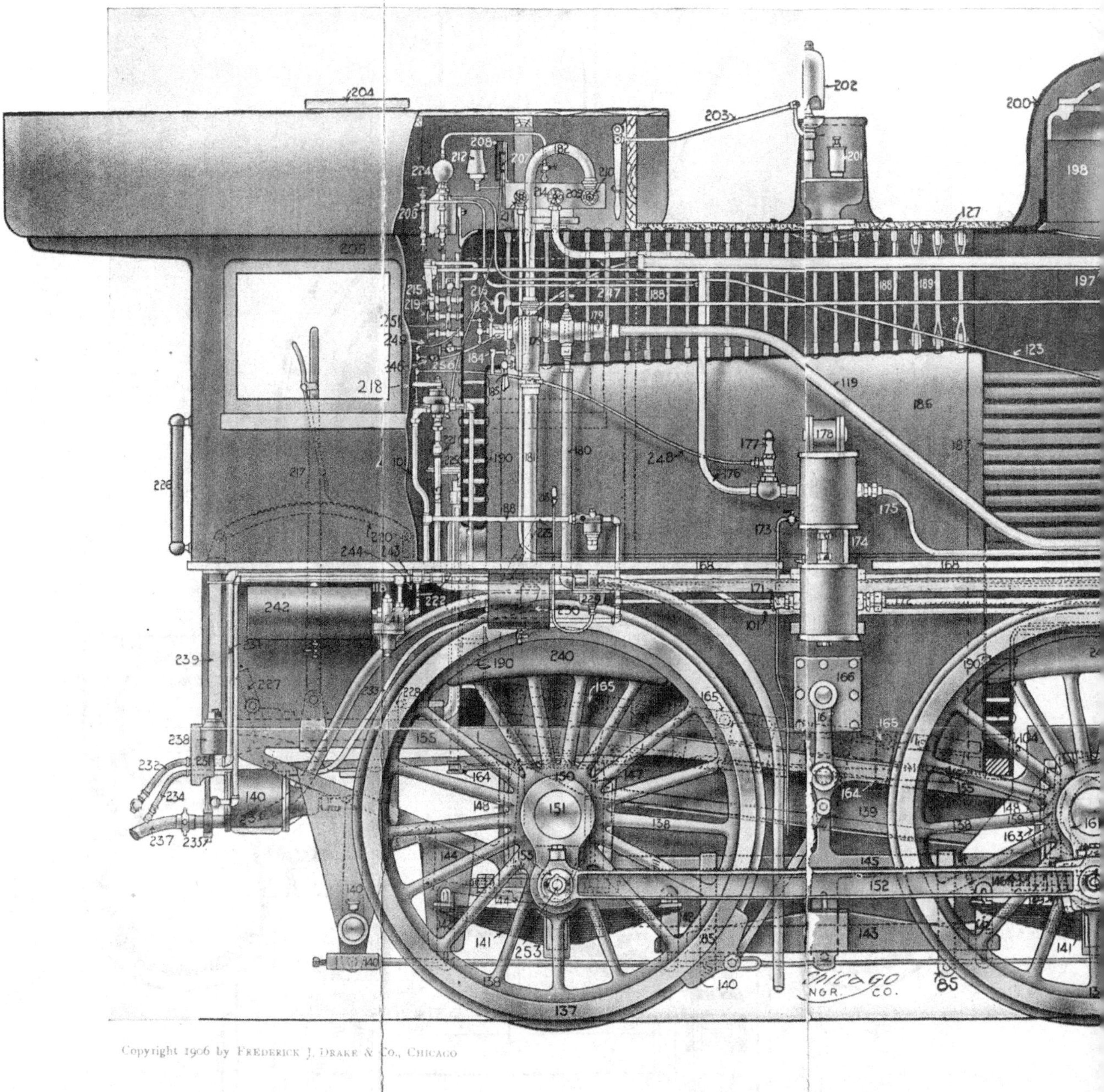

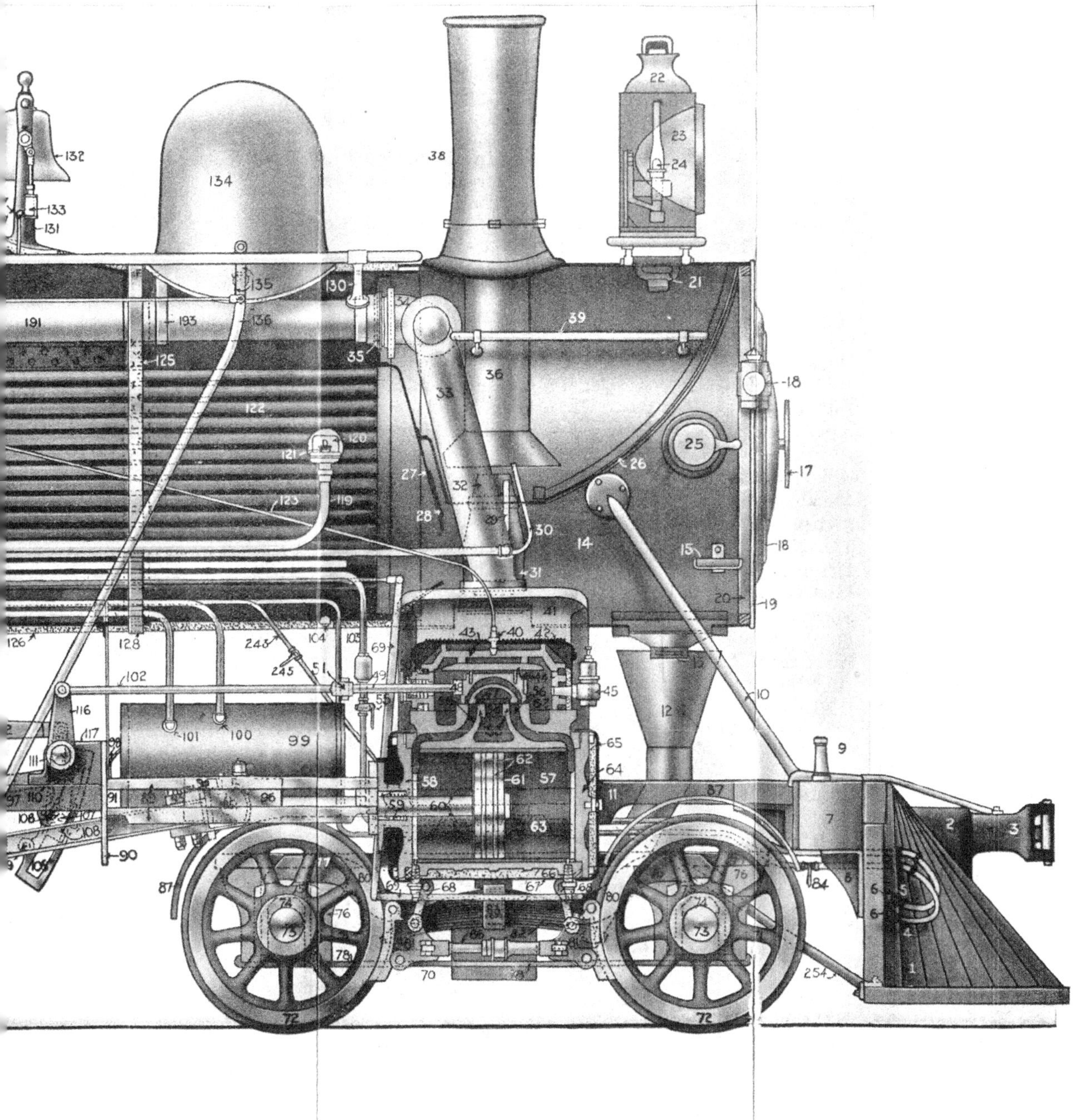

Guide to numbers on previous page spread.

1 Pilot.
2 Draw Bar Plate.
3 Coupler.
4 Air Signal Hose.
5 Air Brake Hose.
6 Hose Hangers.
7 Buffer Beam.
8 Pilot Bracket.
9 Flagstaff.
10 Arch Brace.
11 Front Frame.
12 Cinder Chute.
13 Cinder Chute Side.
14 Extension Front.
15 Headlight Step.
16 Signal Lamp.
17 Number Plate.
18 Smoke Arch Door.
19 Smoke Arch Front.
20 Smoke Arch Ring.
21 Headlight Bracket.
22 Headlight Case.
23 Headlight Reflector.
24 Headlight Burner.
25 Cleaning Door.
26 Netting.
27 Deflector Plate.
28 Deflector Plate Adjuster.
29 Air Pump Exhaust Pipe.
30 Blower.
31 Nozzle Stand.
32 Nozzle Tip.
33 Steam Pipe. (2)
34 T or Nigger Head.
35 Dry Pipe Joint.
36 Petticoat or Draft Pipe.
37 Stack Base.
38 Smoke Stack.
39 Arch Hand Rail.
40 Oil Pipe Plug.
41 Cylinder Saddle.
42 Steam Chest Casing Cover.
43 Steam Chest Cover.
44 Steam Chest.
45 Relief Valve.
46 Balance Plate.
47 Balance Slide Valve.
48 Valve Yoke.
49 Valve Stem.
50 Valve Stem Packing.
51 Valve Stem Connection.
52 Valve Seat.
53 Bridges.
54 Exhaust Port.
55 Front Train Line Cock.
56 Steam Ports.
57 Cylinder.
58 Back Cylinder-Head.
59 Piston Packing.
60 Piston Rod.
61 Piston Head.
62 Piston Packing Rings.
63 Truck Center Casting.
64 Front Cylinder-Head.
65 Cylinder-Head Casing.
66 Cylinder Lagging.
67 Cylinder Casing.
68 Cylinder Cocks.
69 Cylinder Cocks Rigging.
70 Engine Trucks.
71 Engine Truck Wheel.
72 Engine Truck Tire.
73 Engine Truck Axle.
74 Engine Truck Brass.
75 Engine Truck Box.
76 Engine Truck Pedestal.
77 Engine Truck Frame.
78 Engine Truck Pedestal Brace.
79 Engine Truck Frame Brace.
80 Engine Truck Equalizer
81 Engine Truck Spring Hanger.
82 Engine Truck Spring.
83 Engine Truck Spring Band.
84 Front Signal Line Cock.
85 Safety Hanger.
86 Truck Brake.
87 Wheel Guard.
88 Air Signal Pipe.
89 Guides.
90 Guide Yoke.
91 Guide Block.
92 Main Rod.
93 Main Rod Front Strap.
94 Key.
95 Crosshead Pin.
96 Crosshead.
97 Front Frame.
98 Air Drum Bracket.
99 Air Drum.
100 Pump Connection.
101 Train Pipe Connection from Main Reservoir.
102 Valve Stem Rod.
103 Train Pipe.
104 Wash Out Plugs.
105 Link.
106 Suspension Stud.
107 Link Block Pin.
108 Link Block.
109 Eccentric Connection, Back Up.
110 Eccentric Connection, Go Ahead.
111 Link Hanger.
112 Tumbling Shaft Arm.
113 Tumbling Shaft.
114 Tumbling Shaft Lever.
115 Counter Balance Spring and Rig.
116 Rocker.
117 Rocker Box.
118 Reach Rod.
119 Branch Pipe.
120 Check Valve Case.
121 Check Valve.
122 Flues.
123 Oil Pipe.
124 Horizontal Boiler Seam
125 Circumferencial Seam.
126 Boiler Lagging.
127 Boiler Jacket.

128 Jacket Bands.
129 Hand Rail.
130 Hand Rail Bracket.
131 Bell Stand.
132 Bell.
133 Air Bell Ringer.
134 Sand Box.
135 Sand Box Lever.
136 Sand Pipe.
137 Driving Wheel Tire.
138 Driving Wheel Centers.
139 Ash Pan.
140 Driver Brakes.
141 Driver Springs.
142 Driver Spring Hangers.
143 Driver Spring Equalizers.
144 Driver Spring Hanger Brace.
145 Lower Rail of Frame.
146 Pedestal Brace.
147 Driving Box Shoe.
148 Driving Box Wedge.
149 Wedge Bolt.
150 Driving Box.
151 Driving Axle.
152 Side or Parallel Rod.
153 Rod Bush.
154 Main Rod Connection.
155 Main Frame.
156 Frame Brace.
157 Frame Splice.
158 Go Ahead Eccentric.
159 Back Up Eccentric.
160 Go Ahead Eccentric Rod.
161 Go Ahead Eccentric Strap.
162 Back Up Eccentric Rod.
163 Back Up Eccentric Strap.
164 Grate Shaking Rig.
165 Rocking Grates.
166 Expansion Pad.
167 Expansion Link.
168 Running Board.
169 Air Cylinder Brake Pump.
170 Steam Cylinder Brake Pump.
171 Air Strainer.
172 Delivery to Drum.
173 Drip Cock.
174 Pump Piston Packing.
175 Pump Exhaust Connection.
176 Pump Steam Connection.
177 Governor.
178 Pump Valve Case.
179 Injector.
180 Injector Overflow.
181 Water Pipe.
182 Steam Pipe.
183 Steam Valve.
184 Primer.
185 Water Valve.
186 Fire Box.
187 Tube Sheet.
188 Radial Stay Bolts.
189 Sling Stay.
190 Stay Bolts.
191 Dry Pipe.
192 Stand Pipe.
193 Dry Pipe Hangers.
194 Throttle Pipe.
195 Throttle Valve.
196 Throttle Bell Crank.
197 Throttle Stem.
198 Dome.
199 Dome Cap.
200 Dome Casing.
201 Safety Valves.
202 Chime Whistles.
203 Whistle Rig.
204 Ventilator.
205 Cab.
206 Air Pump Lubricator.
207 Air Gauge.
208 Steam Gauge.
209 Steam Turret.
210 Injector Throttle.
211 Blower Cock.
212 Gauge Lamp.
213 Signal Whistle.
214 Air Pump Throttle.
215 Throttle Lever.
216 Sand Lever.
217 Reverse Lever.
218 Engineer's Brake Valve.
219 Gauge Cocks.
220 Quadrant.
221 Cut Out Valve.
222 Fire Door.
223 Cylinder Cock Lever.
224 Sight-Feed Lubricator.
225 Oil Can Shelf.
226 Hand Hold.
227 Shake Lever Stub.
228 Ash Pan Damper Handle.
229 Whistle Signal Valve.
230 Brake Valve Reservoir.
231 Train Pipe.
232 Train Pipe Hose.
233 Signal Pipe.
234 Signal Pipe Hose.
235 Feed Pipe Hanger.
236 Feed Pipe.
237 Feed Pipe Hose.
238 Tall Piece of Frame.
239 Cab Bracket.
240 Counter Balance Weight.
241 Engine Brake Triple Valve.
242 Engine Brake Auxiliary.
243 Connection to Truck Brake Cylinder.
244 Driver Brake Cut Out Cock.
245 Truck Brake Cut Out Cock.
246 Bell Ringer Valve.
247 Air Pipe to Bell Ringer.
248 Air Pipe to Governor.
249 Main Reservoir Connection to Air Gauge.
250 Train Line Connection to Air Gauge.
251 Glass Water Gauge.
252 Main Crank Pin.
253 Back Crank Pin.
254 Pilot Brace.

1026

TABLE OF CONTENTS

INTRODUCTION

THE Art of Railroading, or the Technique of Modern Transportation, is designed to meet evident needs in a clearly stated, plain and simple manner. Changed conditions in railway operation make study a necessity. Men must learn in advance of promotion by practical experience, observation and study. Their knowledge, comprehension or understanding of the subject of Locomotive Engineering may be termed "the what," but the art is to know "the how" of this, as of any other subject. To know the art serves to facilitate the performance of duties in the most efficient manner.

The educational requirements and technical skill and ability of railway men is of a constantly higher standard with an ever increasing demand for men of still broader general knowledge and increased technical education.

It is believed The Prior Self-Educational Railway Series will enable those with limited practical experience to more quickly understand the "what" or knowledge of the subject; and will aid the more experienced to make the best practical application of knowledge gained by observation, experience and study.

To those unable to take a technical college course this series will be of inestimable value.

AMERICAN LOCOMOTIVE COMPANY

ENGINEERING DEPARTMENT

CLASSIFICATION OF LOCOMOTIVES

(WHYTE'S SYSTEM)

040	4 WHEEL SWITCHER
060	6 " "
0660	ARTICULATED
080	8 WHEEL SWITCHER
240	4 COUPLED
260	MOGUL
280	CONSOLIDATION
2100	DECAPOD
440	8 WHEEL
460	10 "
480	12 "
042	4 COUPLED & TRAILING
062	6 " "
082	8 " "
044	FORNEY 4 COUPLED
064	" 6 "
046	" 4 "
066	" 6 "
242	COLUMBIA
262	PRAIRIE
282	8 COUPLED DOUBLE ENDER
2102	10 " " "
244	4 " " "
264	6 " " "
284	8 " " "
246	4 " " "
266	6 " " "
442	ATLANTIC
462	PACIFIC
444	4 COUPLED DOUBLE ENDER
464	6 " " "
446	4 " " "
466	6 " " "

The locomotive classification adopted by the American Locomotive Company is based on the representation by numerals of the number and arrangement of the wheels commencing at the front. Thus 260 means a Mogul and 460 a ten wheel engine, the cipher denoting that no trailing truck is used.

The total weight is expressed in 1,000 of pounds. Thus an Atlantic locomotive weighing 176,000 pounds would be classified as a 442-176 type. If the engine is Compound the letter C should be substituted for the dash thus 442 C 176. If tanks are used in place of a separate tender the letter T should be used in place of the dash. Thus a double end suburban locomotive with two wheeled leading truck, six drivers and six wheeled rear truck, weighing 214,000 pounds, would be a 266 T 214 type.

CHAPTER I

FIREMAN'S DUTIES

One of the most important duties of a fireman is to form the habit of being "on time," if possible. He should be on his engine at least thirty minutes before the engine leaves the house. He will then have time to get everything in good shape.

First see that the water supply is right, then the coal wet down, cab swept out and windows cleaned, oil cans all filled and in their places, and lamps cleaned and filled with oil. He should also be sure that all needed supplies, such as flags, lanterns, torpedoes, waste, etc., are on hand, and of the right kind.

Another important point, and one in which he is particularly interested, is that the engine is supplied with the proper fire tools—clinker bar, ash pan hoe, slice bar, and such other tools as are needed for the proper manipulation of a fire.

It is the duty of the roundhouse men to see that the sand-box is filled with clean, dry sand, but it is well enough for the fireman to have an eye to that also.

For the beginner, especially, there are a great many details to be learned, and he should get in touch with the engineer as soon as possible and keep in touch with him. In fact, the engineer and fireman should always work together, and strive to be of mutual help to each other in every possible way.

After getting the engine out of the roundhouse and before starting to take her around to the train, he should note carefully that all the switches that he will pass are properly lined up and that the track is clear.

The engine bell should always be rung before starting, and be kept ringing while the engine is moving through the yards. Before starting from a terminal station the fireman should carefully prepare his fire—see that it is burning brightly and that it is heavy enough to prevent the exhaust from pulling it out of the fire-box when starting out. The depth of fire that should be carried on the grate bars depends upon the kind of fuel to be used. If soft coal is the fuel, a fire ten to twelve inches deep should be carried. If hard coal is used, the fire should not be so deep.

Before leaving a terminal the fireman should carefully read the train orders and be certain that he understands them thoroughly. Out on the road he should use his eyes in watching the steam gauge and water glass, also try to familiarize himself with the grades and hills. Some engines steam better with the fire a little deeper along the sides and in the corners of the fire-box, allowing the center of the fire to be more shallow. If there is a brick arch or a water table in the fire-box, care should be taken that plenty of space be maintained between it and the fire. At the beginning of the run the fire is clean, and may be kept a little deeper without danger of clogging.

While the engineer is pulling out of a station and working her up to speed, the fireman should watch his fire closely and keep adding a good supply of coal, as there is danger of the fire being broken by the sharp, heavy exhaust. After a good rate of speed has been attained and the engineer has hooked his reverse lever back, the coal should be added to the fire often and in small quantities at a time, two scoopfuls at each fire being sufficient, always waiting until the black smoke emitted from the stack disappears or at least changes to a light gray color before throwing in a fresh fire, and then placing the coal in the brightest spots. If the train is light, one shovelful at a fire is enough. No set of rules for firing can be laid down that will apply to all conditions. The best rule, especially for a man new in the service, is to always be ready to receive suggestions from the engineer, who has passed through all the various phases of a fireman's apprenticeship and knows, or at least ought to know, his engine thoroughly and how to get the best service out of her. Therefore the fireman should always work under the instructions of the engineer; in fact, never do anything while on duty without first knowing that it would meet with his approval.

Care should be exercised in the regulation of the ash pan dampers for admitting air under the grates. The fireman should study closely the requirements of his fire in this respect. If too small a volume of air is admitted the fire will not burn as lively as it should, and if too much air enters the fire-box the gases will be chilled. Keep the ash pan clean and the grates will last longer.

As the exhaust is the life breath of the locomotive, it might be well at this point to explain why it creates such a tremendous draft. The reason is, because of the volume and velocity of the steam as it issues from the exhaust nozzles. The air and gases in the stack are carried out or forced out of the stack by the exhaust, and this creates a partial vacuum in the smoke arch, into which the air and gases pass from the fire-box through the flues. Fresh air is also being forced into the fire-box through the grates and other apertures by the atmospheric pressure. The blower operates upon the same principle, although on a much smaller scale.

It may be used to urge the fire when the engine is not working steam. The blower should also be used while cleaning the fire; it will clear the dust and ashes from the flues. If the engine is pulling a passenger train and the engineer is about to make a stop at a station, the fireman should, as soon as the throttle is closed,

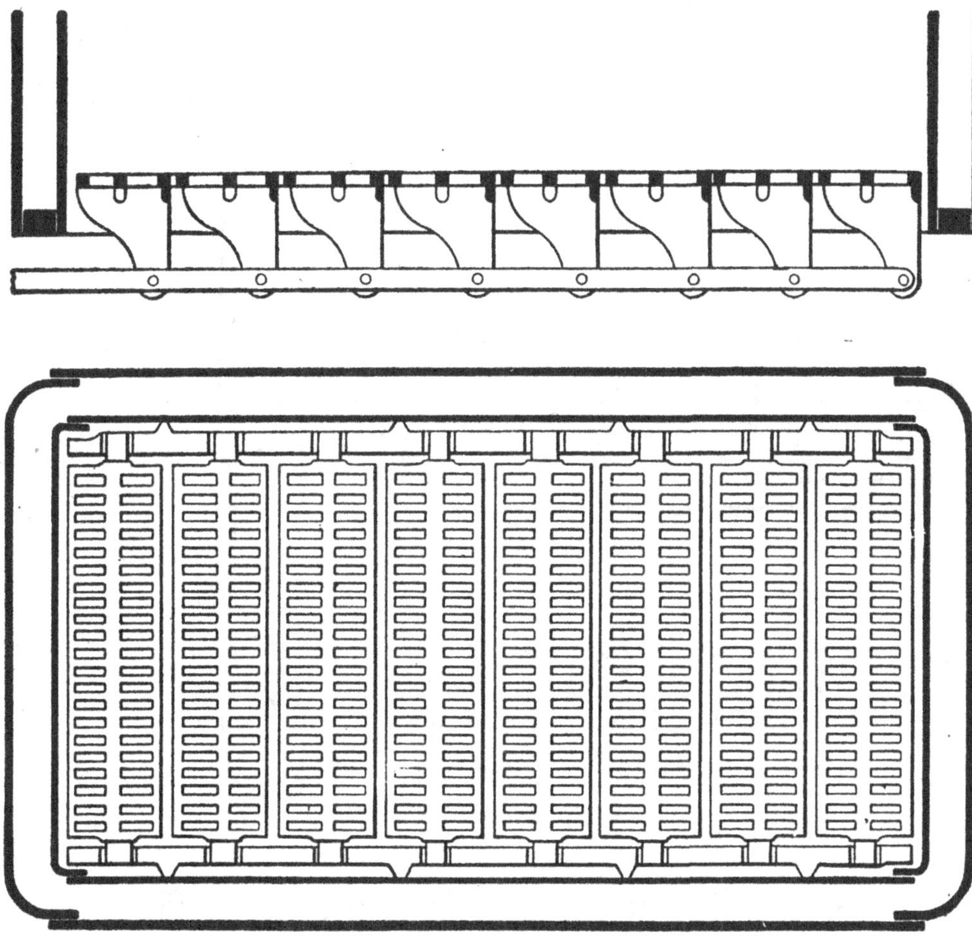

FIGURE 1

put on the blower lightly and open the fire door one-half inch, just sufficient to allow a small volume of air to enter the fire-box above the fire. This will prevent the engine from throwing out a great volume of dense black smoke while making the stop.

As the grate bars are a part of the engine with which

the fireman is particularly interested, a brief description of the various types will be here given. The old-fashioned grate bars for burning wood are too familiar to need describing, being simply plain cast iron stationary bars with narrow slots between them. For soft coal various styles of rocking grates are used. Figs. 1 and 2 show plan and sectional views of rocking grates. The method of shaking is also illustrated in Fig. 2, together with the dump grate at the front to be

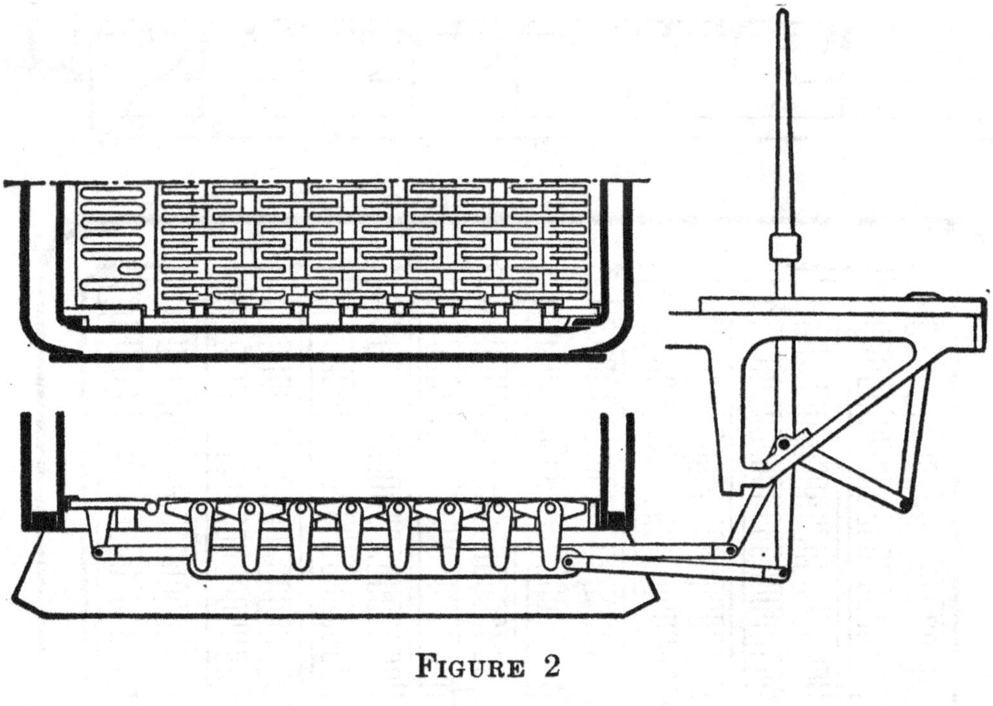

FIGURE 2

used when cleaning the fire. For burning hard coal a larger grate area is required than with soft coal, for the reason that a hard coal fire must be kept more shallow than a soft coal fire. The grate for hard coal is long, and instead of being made of cast iron it consists of horizontal wrought iron water tubes in connection with the water space, thus permitting a free circulation of water through them. This plan not only prevents the grates from burning out, but it also serves to utilize a portion of heat that would otherwise be wasted.

Fig. 3 shows a plan and Fig. 4 an elevation of a set of water grates. Provision is made for drawing or cleaning the fire, by making every fourth or fifth tube solid and allowing it to project clear through both walls of the back end of the fire-box through thimbles inserted for that purpose. These solid tubes have rings on their back ends by which they may be withdrawn, and the front end rests upon a bearing bar.

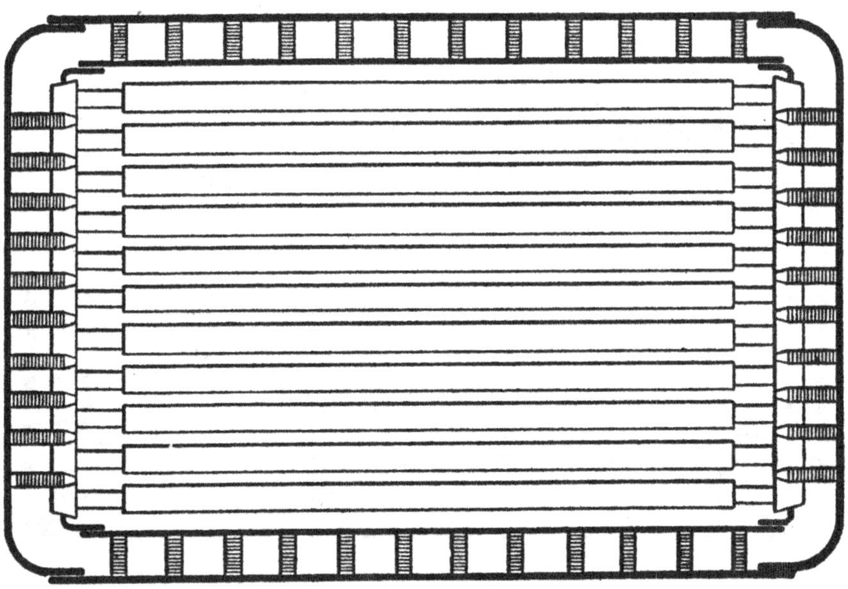

FIGURE 3

The tubes of a water grate are made water-tight by being caulked into the inside sheet at the front and back ends of the fire-box.

About twenty square feet of grate surface is needed to burn one ton of soft coal per hour.

As the steam gauge is an instrument that is particularly interesting to the fireman, it is fitting that a short description of it be inserted here. There are different types of steam gauges in use, but the one most commonly used, and which no doubt is the most reliable, is known as the Bourdon spring gauge. This gauge consists of a thin, curved, flattened metallic tube, closed at both ends and connected to the steam space of the boiler by a small pipe, bent at some portion of its length into a curve or circle that becomes filled with water of condensation, and thus prevents the hot live steam from coming directly in contact with the spring, while at the same time the full pressure of steam in

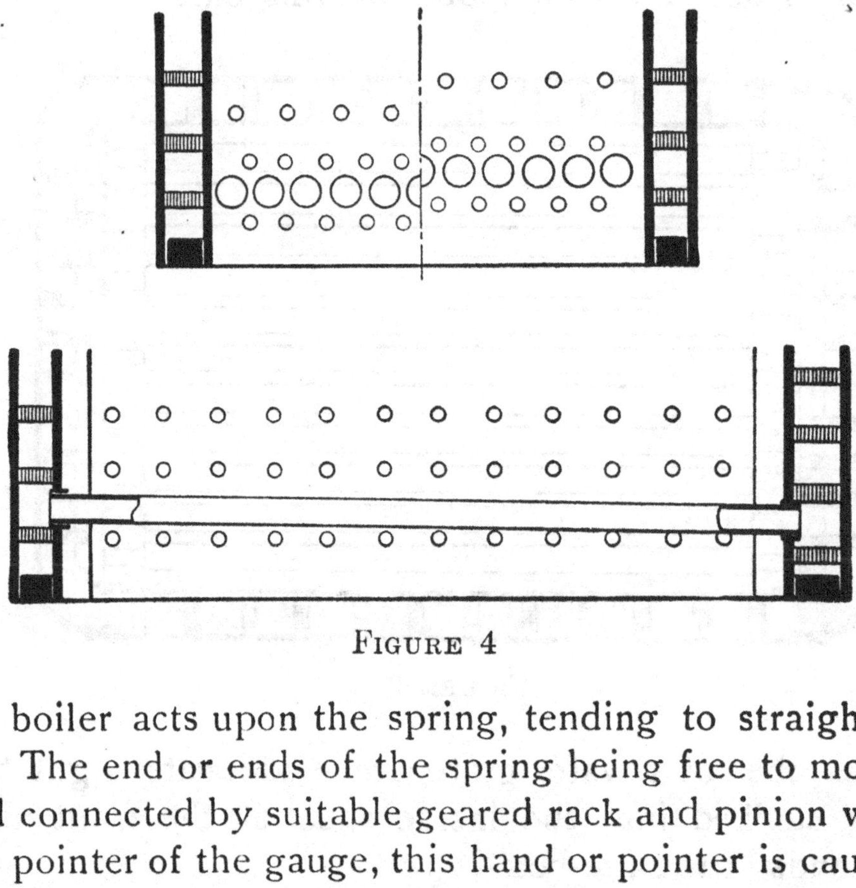

FIGURE 4

the boiler acts upon the spring, tending to straighten it. The end or ends of the spring being free to move, and connected by suitable geared rack and pinion with the pointer of the gauge, this hand or pointer is caused to move across the dial, thus indicating the pressure of steam per square inch in the boiler. When there is no pressure in the boiler the hand should point to 0.

Steam gauges should be tested frequently by comparing them with a test gauge that has been tested against a column of mercury.

The safety valve, or pop valve as it is more familiarly known, is another very important part of a locomotive with which the fireman has business. The object aimed at in equipping a locomotive boiler, or any other boiler, with a safety valve is that the steam pressure may be kept within a safe limit. There should always be two pop valves on a locomotive boiler, so that if one becomes corroded and sticks to its seat the other one will act, thus insuring safety.

The principle of a pop valve's action is this: It is held to its seat by a coil spring that has previously been adjusted to the required amount of resistance. When the pressure under the valve exceeds the resistance of the spring, the valve will rise from its seat and allow the steam to escape until the pressure is lower than the resistance of the spring. The valve will then close at once.

When the fire becomes dull and heavy, caused by ashes accumulating on the grate bars, the grates should be shaken up, which is best done while the engine is working at a moderate speed, or at least when the blower is on. The ash pan should be kept clean and free from ashes. This will allow a free draft of air and prevent the burning out of the grate bars.

The fireman should endeavor while out on the run to keep as even a temperature as possible in the fire-box, and this can only be done by firing light and often, keeping the grates free by shaking and by watching the water level closely. He should have and keep his mind constantly upon his work, always striving to do better to-day than he did yesterday, and his reward is sure to come.

Upon arriving at the end of the run he should take in his flags, or blow out his lamps, and see that the engine has sufficient fire and water to last until the hostler gets around.

One of the duties that a fireman owes to himself, as well as to his employers, is that he utilize his spare moments in the study of the theory of combustion, the composition of coal, the nature of heat, and various other problems connected with the generation of steam. He will be called upon to undergo an examination as to his knowledge of these questions at some stage of his apprenticeship, and the more intelligence he displays and the more thorough his answers, the faster will be his promotion. Therefore the author considers it fitting and proper that a space be given over at this point for the discussion of these important subjects.

Combustion. One of the main factors in the combustion of coal is the proper supply of air. Air is composed of two gases, oxygen and nitrogen, in the proportion, by volume, of 21 per cent of oxygen and 79 per cent of nitrogen, or by weight, 23 per cent of oxygen and 77 per cent of nitrogen.

The composition of pure dry air is as follows:

By volume, 20.91 parts O. and 79.09 parts N.

By weight, 23.15 parts O. and 76.85 parts N.

Air is a mixture and not a chemical combination of these two elements. The principal constituent of coal and most other fuels, whether solid, liquid or gaseous, is carbon. Hydrogen is a light combustible gas and, combined either with carbon or with carbon and oxygen, in various proportions, is also a valuable constituent of fuels, notably of bituminous coal. The heating value of one pound of pure carbon is rated at 14,500 heat units, while one pound of hydrogen gas contains 62,000 heat units.

Analysis of coal shows that it contains moisture, fixed carbon, volatile matter, ash and sulphur in various proportions according to the quality of the coal. The following table will show the composition of the principal bituminous coals in use in this country for steam purposes. Two samples are selected from each of the great coal producing states, with the exception of Illinois, from which four were taken.

TABLE I

State	Kind of Coal	Moisture	Volatile Matter	Fixed Carbon	Ash	Sulphur
Pennsylvania	Youghiogheny	1.03	36.49	59.05	2.61	0.81
"	Connellsville	1.26	30.10	59.61	8.23	0.78
West Virginia	Quinimont	0.76	18.65	79.26	1.11	0.23
"	Fire Creek	0.61	22.34	75.02	1.47	0.56
E. Kentucky	Peach Orchard	4.60	35.70	53.28	6.42	1.08
"	Pike County	1.80	26.80	67.60	3.80	0.97
Alabama	Cahaba	1.66	33.28	63.04	2.02	0.53
"	Pratt Co.'s	1.47	32.29	59.50	6.73	1.22
Ohio	Hocking Valley	6.59	35.77	49.64	8.00	1.59
"	Muskingum "	3.47	37.88	53.30	5.35	2.24
Indiana	Block	8.50	31.00	57.50	3.00	
"	"	2.50	44.75	51.25	1.50	
W. Kentucky	Nolin River	4.70	33.24	54.94	11.70	2.54
"	Ohio County	3.70	30.70	45.00	3.16	1.24
Illinois	Big Muddy	6.40	30.60	54.60	8.30	1.50
"	Wilmington	15.50	32.80	39.90	11.80	
"	" screenings	14.00	26.00	34.20	23.80	
"	Duquoin	8.90	23.50	60.60	7.00	

The process of combustion consists in the union of the carbon and hydrogen of the fuel with the oxygen of the air. Each atom of carbon combines with two atoms of oxygen, and the energetic vibration set up by their combination is heat. Bituminous coal contains a large percentage of volatile matter which is released and flashes into flame when the coal is thrown into the furnace, and unless air is supplied in large amounts at this stage of the combustion there will be an excess of smoke and consequent loss of carbon. On the other hand, there is a loss in admitting too much air, because the surplus is heated to the temperature of the furnace without aiding the combustion and will carry off to the stack just as many heat units as were required to raise it from the temperature at which it entered the fire-box to that at which it leaves the flues. Some kinds of coal need more air for their combustion than do others, and good judgment and close observation are needed on the part of the fireman to properly regulate the supply.

The quantity of air required for the combustion of one pound of coal is, by volume, about 150 cu. ft.; by weight, about 12 lbs.

The temperature of the fire-box is usually about 2500°, in some cases reaching as high as 3000°. The temperature of the escaping gases should not be much above nor below 400° F. for bituminous coal.

In order to attain the highest economy in the burning of coal in boiler furnaces two factors are indispensable, viz., a constant high furnace temperature and quick combustion, and these factors can only be secured by supplying the fresh coal constantly just as fast as it is burned, and also by preventing as much as possible the admission of cold air at the furnace. The nitrogen in the atmosphere does not promote combustion, but it enters the fire-box along with the oxygen, and the heat required to raise its temperature to that of the other gases is practically wasted, and as has already been explained, if a surplus of cold air is allowed to pass into the fire-box the waste of heat becomes still greater.

Heat. All matter, whether solid, liquid, or gaseous, consists of molecules or atoms, which are in a state of continual vibration, and the result of this vibration is heat. The intensity of the heat evolved depends upon the degree of agitation to which the molecules are subject. Until as late as the beginning of the nineteenth century two rival theories in regard to the nature of heat had been advocated by scientists. The older of these theories was that heat was a material substance, a subtle elastic fluid termed caloric, and that this fluid penetrated matter as water penetrates a sponge. But this theory was shown to be false by the wonderful researches and experiments of Count Rumford at Munich, Bavaria, in 1798.

By means of the friction between two heavy metallic bodies placed in a wooden trough filled with water, one of the pieces of metal being rotated by machinery driven by horses, Count Rumford succeeded in raising the temperature of the water in two and one-half hours from its original temperature of 60° to 212° F., the boiling point, thus demonstrating that heat is not a material substance, but that it is due to vibration or motion, an internal commotion among the molecules of matter. This theory, known as the Kinetic theory of heat, has since been generally accepted, although it was nearly fifty years after Rumford advocated it in a paper read before the Royal Society of Great Britain in 1798, before scientists generally became converted to this idea of the nature of heat, and the science of Thermo Dynamics was placed on a firm basis.

During the period from 1840 to 1849 Dr. Joule made a series of experiments which not only confirmed the truth of Count Rumford's theory that heat was not a material substance but a form of energy which may be applied to or taken away from bodies, but Joule's experiments also established a method of estimating in mechanical units or foot pounds the amount of that energy. This latter was a most important discovery, because by means of it the exact relation between heat and work can be accurately measured.

The first law of thermo dynamics is this: Heat and mechanical energy or work are mutually convertible. That is, a certain amount of work will produce a certain amount of heat, and the heat thus produced is capable of producing by its disappearance a fixed amount of mechanical energy if rightly applied. The mechanical energy in the form of heat which, through the medium of the steam engine, has revolutionized the world, was first stored up by the sun's heat millions of years ago in the coal, which in turn, by combustion, is made to release it for purposes of mechanical work.

The general principles of Dr. Joule's device for measuring the amount of work in heat are illustrated in Fig. 5. It consisted of a small copper cylinder containing a known quantity of water at a known temperature. Inside the cylinder and extending through the top was a vertical shaft to which were fixed paddles for stirring the water. Stationary vanes were also placed inside the cylinder. Motion was imparted to the shaft through the medium of a cord or small rope coiled around a drum near the top of the shaft and running over a grooved pulley or sheave. To the free end of the cord a known weight was attached. This weight was allowed to fall through a certain distance, and in falling it turned the shaft with its paddles, which in turn agitated the water, thus producing a certain amount of heat. To illustrate, suppose the weight to be 77.8 lbs., and that by means of the crank at the top end of the shaft it has been raised to the zero mark at the top of the scale. (See Fig. 5.) One pound of water at 39.1° F. is poured into the copper cylinder,

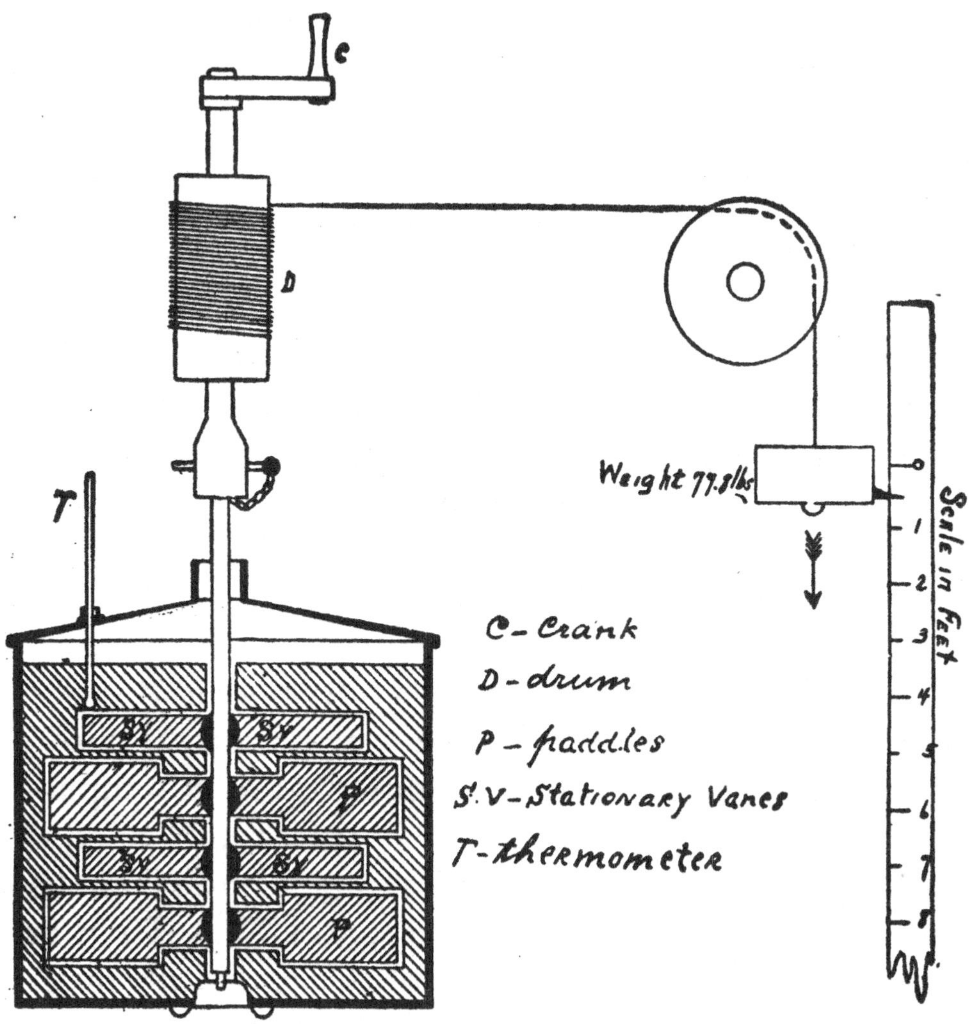

FIGURE 5

which is then closed and the weight released. At the moment the weight passes the 10 ft. mark on the scale the thermometer attached to the cylinder will indicate that the temperature of the water has been raised one degree. Then multiplying the number of pounds in the weight by the distance in feet through which it fell will give the number of foot pounds of work done. Thus, 77.8 lbs. × 10 ft. = 778 foot pounds.

The heat unit or British thermal unit (B. T. U.) is the quantity of heat required to raise the temperature of one pound of water one degree, or from 39° to 40° F., and the amount of mechanical work required to produce a unit of heat is 778 foot pounds. Therefore the mechanical equivalent of heat is the energy required to raise 778 lbs. one foot high, or 77.8 lbs. 10 ft. high, or 1 lb. 778 feet high. Or again, suppose a one-pound weight falls through a space of 778 ft. or a weight of 778 lbs. falls one foot, enough mechanical energy would thus be developed to raise a pound of water one degree in temperature, provided all the energy so developed could be utilized in churning or stirring the water, as in Joule's machine. Hence the mechanical equivalent of heat is 778 foot pounds.

Specific Heat. The specific heat of any substance is the ratio of the quantity of heat required to raise a given weight of that substance one degree in temperature to the quantity of heat required to raise an equal weight of water one degree in temperature when the water is at its maximum density, 39.1° F. To illustrate, take the specific heat of lead, for instance, which is .031, while the specific heat of water is 1. That means that it would require 31 times as much heat to raise one pound of water one degree in temperature as it would to raise the temperature of a pound of lead one degree.

The following table gives the specific heat of different substances in which engineers are most generally interested.

TABLE 2

Water at 39.1° F.	**1.000**
Ice at 32° F.	.504
Steam at 212° F.	.480
Mercury	.033
Cast iron	.130
Wrought iron	.113
Soft steel	.116
Copper	.095
Lead	.031
Coal	.240
Air	.238
Hydrogen	3.404
Oxygen	.218
Nitrogen	.244

Sensible Heat and Latent Heat. The plainest and most simple definition of these two terms is that given by Sir Wm. Thomson. He says: "Heat given to a body and warming it is sensible heat. Heat given to a body and not warming it is latent heat." Sensible heat in a substance is the heat that can be measured in degrees of a thermometer, while latent heat is the heat in any substance that is not shown by the thermometer.

To illustrate this more fully, a brief reference to some experiments made by Professor Black in 1762 will no doubt make the matter plain. It will be remembered that at that early date comparatively little was known of the true nature of heat; hence Professor Black's investigations and discoveries along this line appear all the more wonderful. He procured equal weights of ice at 32° F. and water at the same temperature, that is, just at the freezing point, and placing them in separate glass vessels, suspended the vessels in a room in which the uniform temperature was 47° F. He noticed that in one-half hour the water had increased 7° F. in temperature, but that twenty half hours elapsed before all of the ice was melted. Therefore he reasoned that twenty times more heat had entered the ice than had entered the water, because at the end of the twenty half hours, when the ice was all melted, the water in both vessels was of the same temperature. The water, having absorbed 7° of heat during the first half hour, must have continued to absorb heat at the same rate during the whole of the twenty half hours, although the thermometer did not indicate it. From this he calculated that 7° × 20 = 140° of heat had become latent or hidden in the water.

In another experiment Professor Black placed a lump of melting ice, which he estimated to be at a temperature of 33° F. on the surface, in a vessel containing the same weight of water at 176° F., and he observed that when the whole of the ice had been melted the temperature of the water was 33° F., thus proving that 143° of heat (176° – 33°) had been absorbed in melting the ice and was at that moment latent in the water. By these two experiments Professor Black established the theory of the latent heat of water, and his estimate was very near the truth, because the results obtained since that time by the greatest experimenters show that the latent heat of water is 142 heat units, or B. T. U.

Black's experiment for ascertaining the latent heat in steam at atmospheric pressure was made in the following simple manner: He placed a flat, open tin dish on a hot plate over a fire and into the dish he put a small quantity of water at 50° F. In four minutes the water began to boil, and in twenty minutes more it had all evaporated. In the first four minutes the temperature had increased 212° – 50° = 162°, and the temperature remained at 212° throughout the twenty minutes that it required to evaporate all the water, despite the fact that the water had been receiving heat during this period at the same rate as during the first four minutes. He therefore reasoned that in the twenty minutes the water had absorbed five times as much heat as it had in the four minutes, or $162° \times 5 = 810°$, without any sensible rise in temperature. Therefore the 810° became latent in the steam. Owing to the crude nature of the experiment Professor Black's estimate of the number of degrees of latent heat in steam was incorrect, as it has been proven by many famous experimenters since then that the latent heat of steam at atmospheric pressure is 965.7 B. T. U.

It will thus be perceived that what is meant by the term latent heat is that quantity of heat which becomes hidden or latent when the state of a body is changed from a solid to a liquid, as in the case of melting ice, or from a liquid to a gaseous state, as with water evaporated into steam. But the heat so disappearing has not been lost; on the contrary it has, while becoming latent, been doing an immense amount of work, as can easily be ascertained by means of a few simple figures. It has been seen that a heat unit is the quantity of heat required to raise one pound of water one degree in temperature and also that the mechanical equivalent of heat, or, in other words, the mechanical energy stored in one heat unit, is equal to 778 foot pounds of work.

A horse power equals 33,000 ft. lbs. of energy in one minute of time, and a heat unit $= 778 \div 33{,}000 = .0236$, or about $\frac{1}{43}$ of a horse power. The work done by the heat which becomes latent in converting one pound of ice at 32° F. into water at the same temperature = 142 heat units × 778 ft. lbs. = 110,476 ft. lbs., which divided by 33,000 equals 3.34 horse power. Again, by the evaporation of one pound of water from

32° F. into steam at atmospheric pressure, 965.7 units of heat become latent in the steam and the work done = 965.7 × 778 = 751,314 ft. lbs. = 22.7 horse power. It will thus be seen what tremendous energy lies stored in one pound of coal, which contains from 12,000 to 14,500 heat units, provided all the heat could be utilized in an engine.

Total Heat of Evaporation. In order to raise the temperature of one pound of water from the freezing point, 32° F., to the boiling point, 212° F., there must be added to the temperature of the water 212° − 32° = 180°. This represents the sensible heat. Then to make the water boil at atmospheric pressure, or, in other words, to evaporate it, there must still be added 965.7 B. T. U., thus 180 + 965.7 = 1,145.7, or in round numbers 1,146 heat units. This represents what is termed the total heat of evaporation at atmospheric pressure and is the sum of the sensible and latent heat in steam at that pressure. But if a thermometer were held in steam evaporating into the open air, as, for instance, in front of the spout of a tea kettle, it would indicate but 212° F.

When steam is generated at a higher pressure than 212°, the sensible heat increases and the latent heat decreases slowly, while at the same time the total heat of evaporation slowly increases as the pressure increases, but not in the same ratio. As, for instance, the total heat in steam at atmospheric pressure is 1,146 B. T. U., while the total heat in steam at 100 lbs. gauge pressure is 1,185 B. T. U., and the sensible temperature of steam at atmospheric pressure is 212°, while at 100 lbs. gauge pressure the temperature is 338 and the latent heat is 876 B. T. U. See Table 4.

Water. The elements that enter into the composition of pure water are the two gases, hydrogen and oxygen, in the following proportions:

By volume,	hydrogen	2,	oxygen	1.
By weight,	"	11.1,	"	88.9.

Perfectly pure water is not attainable, neither is it desirable nor necessary to the welfare of the human race, because the presence of certain proportions of air and ammonia add greatly to its value as an agent for manufacturing purposes and for generating steam. The nearest approach to pure water is rain water, but even this contains 2.5 volumes of air to each 100 volumes of water. Pure distilled water, such for instance as the return water from steam heating systems, is not desirable for use alone in a boiler, as it will cause corrosion and pitting of the sheets, but if it is mixed with other water before going into the boiler its use is highly beneficial, as it will prevent to a certain degree the formation of scale and incrustation. Nearly all water used for the generation of steam in boilers contains more or less scale-forming matter, such as the carbonates of lime and magnesia, the sulphates of lime and magnesia, oxide of iron, silica and organic matter, which latter tends to cause foaming in boilers.

The carbonates of lime and magnesia are the chief causes of incrustation. The sulphate of lime forms a hard crystalline scale which is extremely difficult to remove when once formed on the sheets and tubes of boilers. Of late years the intelligent application of chemistry to the analyzing of feed waters has been of great benefit to engineers and steam users, in that it has enabled them to properly treat the water with solvents either before it is pumped into the boiler or by the introduction into the boiler of certain scale preventing compounds made especially for treating the particular kind of water used. Where it is necessary to treat water in this manner, great care and watchfulness should be exercised by the engineer in the selection and use of a boiler compound.

From 10 to 40 grains of mineral matter per gallon are held in solution by the waters of the different rivers, streams and lakes; well and mine water contain more.

Water contracts and becomes denser in cooling until it reaches a temperature of 39.1° F., its point of greatest density. Below this temperature it expands, and at 32° F. it becomes solid or freezes, and in the act of freezing it expands considerably, as every engineer who has had to deal with frozen water pipes can testify.

Water is 815 times heavier than atmospheric air. The weight of a cubic foot of water at 39.1° is approximately 62.5 lbs., although authorities differ on this matter, some of them placing it at 62.379 lbs., and others at 62.425 lbs. per cubic foot. As its temperature increases its weight per cubic foot decreases, until at 212° F. one cubic foot weighs 59.76 lbs.

The table which follows is compiled from various sources and gives the weight of a cubic foot of water at different temperatures.

TABLE 3

Temperature	Weight per Cubic Foot	Temperature	Weight per Cubic Foot	Temperature	Weight per Cubic Foot
32° F.	62.42 lbs.	132° F.	61.52 lbs.	230° F.	59.37 lbs.
42°	62.42	142°	61.34	240°	59.10
52°	62.40	152°	61.14	250°	58.85
62°	62.36	162°	60.94	260°	58.52
72°	62.30	172°	60.73	270°	58.21
82°	62.21	182°	60.50	300°	57.26
92°	62.11	192°	60.27	330°	56.24
102°	62.00	202°	60.02	360°	55.16
112°	61.86	212°	59.76	390°	54.03
122°	61.70	220°	59.64	420°	52.86

The boiling point of water varies according to the pressure to which it is subject. In the open air at sea level the boiling point is 212° F. When confined in a boiler under steam pressure the boiling point of water depends upon the pressure and temperature of the steam, as, for instance, at 100 lbs. gauge pressure the temperature of the steam is 338° F., to which temperature the water must be raised before its molecules will separate and be converted into steam. In the absence of any pressure, as in a perfect vacuum, water boils at 32° F. temperature. In a vacuum of 28 in., corresponding to an absolute pressure of .943 lbs., water will boil at 100°, and in a vacuum of 26 in., at which the absolute pressure is 2 lbs., the boiling point of water is 127° F. On the tops of high mountains in a rarefied atmosphere water will boil at a much lower temperature than at sea level; for instance, at an altitude of 15,000 ft. above sea level water boils at 184° F.

Steam. Having discussed to some extent the physical properties of water, it is now in order to devote some time to the study of the nature of steam, which is simply water in its gaseous form, made so by the application of heat.

As has been stated in another portion of this book, matter consists of molecules or atoms inconceivably small in size, yet each having an individuality, and in the case of solids or liquids, each having a mutual cohesion or attraction for the other, and all being in a state of continual vibration more or less violent according to the temperature of the body.

The law of gravitation, which holds the universe together, also exerts its wonderful influence on these atoms and causes them to hold together with more or less tenacity according to the nature of the substance.

Thus it is much more difficult to chip off pieces of iron or granite than it is of wood. But in the case of water and other liquids the atoms, while they adhere to each other to a certain extent, still are not so hard to separate; in fact, they are to some extent repulsive to each other, and unless confined within certain bounds the atoms will gradually scatter and spread out, and finally either be evaporated or sink out of sight in the earth's surface. Heat applied to any substance tends to accelerate the vibrations of the molecules, and if enough heat is applied it will reduce the hardest substances to a liquid or gaseous state.

The process of the generation of steam from water is simply an increase of the natural vibrations of the molecules of the water, caused by the application of heat, until they lose all attraction for each other and become instead entirely repulsive, and unless confined will fly off into space. But, being confined, they continually strike against the sides of the containing vessel, thus causing the pressure which steam or any other gas exerts when under confinement.

Of course steam, like other gases, when under pressure is invisible, but the laws governing its action are well known. These laws, especially those relating to the expansion of steam, will be more fully discussed in the chapter on the Indicator. The temperature of steam in contact with the water from which it is generated, as for instance in the ordinary steam boiler, depends upon the pressure under which it is generated. Thus at atmospheric pressure its temperature is 212° F. If the vessel is closed and the pressure increased the temperature of the steam and also that of the water rises.

Saturated Steam. When steam is taken directly

from the boiler to the engine without being superheated, it is termed saturated steam. This does not necessarily imply that it is wet and mixed with spray and moisture.

Superheated Steam. When steam is conducted into or through a vessel or coils of pipe separate from the boiler in which it was generated and is there heated to a higher temperature than that due to its pressure, it is said to be superheated.

Dry Steam. When steam contains no moisture it is said to be dry. Dry steam may be either saturated or superheated.

Wet Steam. When steam contains mist or spray intermingled, it is termed wet steam, although it may have the same temperature as dry saturated steam of the same pressure.

During the further consideration of steam in this book, saturated steam will be mainly under discussion, for the reason that this is the normal condition of steam as used most generally in steam engines.

Total Heat of Steam. The total heat in steam includes the heat required to raise the temperature of the water from 32° F. to the temperature of the steam plus the heat required to evaporate the water at that temperature. This latter heat becomes latent in the steam, and is therefore called the latent heat of steam.

The work done by the heat acting within the mass of water and causing the molecules to rise to the surface is termed by scientists internal work, and the work done in compressing the steam already formed in the boiler or in pushing it against the superincumbent atmosphere, if the vessel be open, is termed external work. There are, therefore, in reality three elements to be taken into consideration in estimating the total heat of steam, but as the heat expended in doing external work is done within the mass itself, it may, for practical purposes, be included in the general term latent heat of steam.

Density of Steam. The expression density of steam means the actual weight in pounds or fractions of a pound avoirdupois of a given volume of steam, as one cubic foot. This is a very important point for young engineers especially to remember, so as not to get the two terms, pounds pressure and pounds weight, mixed, as some are prone to do.

Volume of Steam. By this term is meant the volume as expressed by the number of cubic feet in one pound weight of steam.

Relative Volume of Steam. This expression has reference to the number of volumes of steam produced from one volume of water. Thus the steam produced by the evaporation of one cubic foot of water from 39° F. into steam at atmospheric pressure will occupy a space of 1646 cu. ft., but, as the steam is compressed and the pressure allowed to rise, the relative volume of the steam becomes smaller, as, for instance, at 100 lbs. gauge pressure the steam produced from one cubic foot of water will occupy but 237.6 cu. ft., and if the same steam was compressed to 1,000 lbs. absolute or 985.3 lbs. gauge pressure it would then occupy only 30 cu. ft.

The condition of steam as regards its dryness may be approximately estimated by observing its appearance as it issues from a pet cock or other small opening into the atmosphere. Dry or nearly dry steam containing about 1 per cent of moisture will be transparent close to the orifice through which it issues, and even if it is of a grayish white color it may be estimated to contain not over 2 per cent of moisture.

Steam in its relation to the engine should be considered in the character of a vehicle for transferring the energy, created by the heat, from the boiler to the engine. For this reason all steam drums, headers and pipes should be thoroughly insulated, in order to prevent, as much as possible, the loss of heat or energy by radiation.

Table 4 gives the physical properties of steam, and is convenient for reference.

TABLE 4

PROPERTIES OF SATURATED STEAM

Vacuum Inches of Mercury	Absolute Pressure Lbs. per Sq. Inch	Temp. Degrees F.	Total Heat above 32° F. In the Water h Heat-units	Total Heat above 32° F. In the Steam H Heat-units	Latent Heat H-h Heat units	Relative Volume	Cubic Feet in 1 Lb. Wt. of Steam	Wt. of 1 Cubic Foot of Steam, Lbs.
29 74	.089	32.	0.	1091 7	1091.7	208,080	3333.3	.0003
29.67	.122	40.	8.	1094.1	1086.1	154,330	2472.2	.0004
29.56	.176	50	18.	1097.2	1079.2	107,630	1724.1	.0006
29.40	.254	60.	28.01	1100.2	1072.2	76,370	1223.4	.0008
29.19	.359	70.	38.02	1103.3	1065.3	54,660	875.61	.0011
28.90	.502	80.	48.04	1106.3	1058.3	39,690	635.80	.0016
28.51	.692	90.	58.06	1109.4	1051.3	29,290	469.20	.0021
28.00	.943	100.	68.08	1112.4	1044.4	21,830	349.70	.0028
27.88	1.	102.1	70.09	1113.1	1043.0	20,623	334.23	.0030
25.85	2.	126.3	94.44	1120.5	1026.0	10,730	173.23	.0058
23.83	3.	141.6	109.9	1125.1	1015.3	7,325	118.00	.0085
21.78	4.	153.1	121.4	1128.6	1007.2	5,588	89.80	.0111
19.74	5.	162.3	130.7	1131.4	1000.7	4,530	72.50	.0137
17.70	6.	170.1	138.6	1133.8	995.2	3 816	61.10	.0163
15.67	7.	176.9	145.4	1135.9	990.5	3,302	53.00	.0189
13.63	8.	182.9	151.5	1137.7	986.2	2,912	46.60	.0214
11.60	9.	188.3	156.9	1139.4	982.4	2,607	41.82	.0239
9.56	10.	193.2	161.9	1140.9	979.0	2,361	37.80	.0264
7.52	11.	197.8	166.5	1142.3	975.8	2,159	34.61	.0289
5.49	12.	202.0	170.7	1143.5	972.8	1,990	31.90	.0314
3.45	13.	205.9	174.7	1144.7	970.0	1,846	29.60	.0338
1.41	14.	209.6	178.4	1145.9	967.4	1,721	27.50	.0363
0.00	14.7	212.0	180.9	1146.0	965.7	1,646	26.36	.0379

TABLE 4—*Continued*

Gauge Pressure Lbs. per Sq. In.	Absolute Pressure Lbs. per Sq. In.	Temp. Degrees F.	Total Heat Above 32° F.		Latent Heat H-h Heat-units	Relative Volume	Cubic Feet in 1 Lb. Wt. of Steam	Wt. of 1 Cubic Foot of Steam, Lbs.
			In the Water h Heat-units	In the Steam H Heat-units				
0.3	15	213.3	181.9	1146.9	965.0	1,614	25.90	.0387
1.3	16	216.3	185.3	1147.9	962.7	1,519	24.33	.0411
2.3	17	219.4	188.4	1148.9	960.5	1,434	23.00	.0435
3.3	18	222.4	191.4	1149.8	958.3	1,359	21.80	.0459
4.3	19	225.2	194.3	1150.6	956.3	1,292	20.70	.0483
5.3	20	227.9	197.0	1151.5	954.4	1,231	19.72	.0507
6.3	21	230.5	199.7	1152.2	952.6	1,176	18.84	.0531
7.3	22	233.0	202.2	1153.0	950.8	1,126	18.03	.0555
8.3	23	235.4	204.7	1153.7	949.1	1,080	17.30	.0578
9.3	24	237.8	207.0	1154.5	947.4	1,038	16.62	.0602
10.3	25	240.0	209.3	1155.1	945.8	998	16.00	.0625
11.3	26	242.2	211.5	1155.8	944.3	962	15.42	.0649
12.3	27	244.3	213.7	1156.4	942.8	929	14.90	.0672
13.3	28	246.3	215.7	1157.1	941.3	898	14.40	.0696
14.3	29	248.3	217.8	1157.7	939.9	869	13.91	.0719
15.3	30	250.2	219.7	1158.3	938.9	841	13.50	.0742
16.3	31	252.1	221.6	1158.8	937.2	816	13.07	.0765
17.3	32	254.0	223.5	1159.4	935.9	792	12.68	.0788
18.3	33	255.7	225.3	1159.9	934.6	769	12.32	.0812
19.3	34	257.5	227.1	1160.5	933.4	748	12.00	.0835
20.3	35	259.2	228.8	1161.0	932.2	728	11.66	.0858
21.3	36	260.8	230.5	1161.5	931.0	709	11.36	.0880
22.3	37	262.5	232.1	1162.0	929.8	691	11.07	.0903
23.3	38	264.0	233.8	1162.5	928.7	674	10.80	.0926
24.3	39	265.6	235.4	1162.9	927.6	658	10.53	.0949
25.3	40	267.1	236.9	1163.4	926.5	642	10.28	.0972
26.3	41	268.6	238.5	1163.9	925.4	627	10.05	.0995
27.3	42	270.1	240.0	1164.3	924.4	613	9.83	.1018
28.3	43	271.5	241.4	1164.7	923.3	600	9.61	.1040
29.3	44	272.9	242.9	1165.2	922.3	587	9.41	.1063
30.3	45	274.3	244.3	1165.6	921.3	575	9.21	.1086
31.3	46	275.7	245.7	1166.0	920.4	563	9.02	.1108
32.3	47	277.0	247.0	1166.4	919.4	552	8.84	.1131
33.3	48	278.3	248.4	1166.8	918.5	541	8.67	.1153
34.3	49	279.6	249.7	1167.2	917.5	531	8.50	.1176
35.3	50	280.9	251.0	1167.6	916.6	520	8.34	.1198
36.3	51	282.1	252.2	1168.0	915.7	511	8.19	.1221
37.3	52	283.3	253.5	1168.4	914.9	502	8.04	.1243

TABLE 4—*Continued*

Gauge Pressure Lbs. per Sq. In.	Absolute Pressure Lbs. per Sq. In.	Temp. Degrees F.	Total Heat above 32° F.		Latent Heat H-h Heat-units	Relative Volume	Cubic Feet in 1 Lb. Wt. of Steam	Wt. of 1 Cubic Foot of Steam, Lbs.
			In the Water h Heat-units	In the Steam H Heat-units				
38.3	53	284.5	254.7	1168.7	914.0	492	7.90	.1266
39.3	54	285.7	256.0	1169.1	913.1	484	7.76	.1288
40.3	55	286.9	257.2	1169.4	912.3	476	7.63	.1311
41.3	56	288.1	258.3	1169.8	911.5	468	7.50	.1333
42.3	57	289.1	259.5	1170.1	910.6	460	7.38	.1355
43.3	58	290.3	260.7	1170.5	909.8	453	7.26	.1377
44.3	59	291.4	261.8	1170.8	909.0	446	7.14	.1400
45.3	60	292.5	262.9	1171.2	908.2	439	7.03	.1422
46.3	61	293.6	264.0	1171.5	907.5	432	6.92	.1444
47.3	62	294.7	265.1	1171.8	906.7	425	6.82	.1466
48.3	63	295.7	266.2	1172.1	905.9	419	6.72	.1488
49.3	64	296.8	267.2	1172.4	905.2	413	6.62	.1511
50.3	65	297.8	268.3	1172.8	904.5	407	6.53	.1533
51.3	66	298.8	269.3	1173.1	903.7	401	6.43	.1555
52.3	67	299.8	270.4	1173.4	903.0	395	6.34	.1577
53 3	68	300.8	271.4	1173.7	902.3	390	6.25	.1599
54.3	69	301.8	272.4	1174.0	901.6	384	6.17	.1621
55.3	70	302.7	273.4	1174.3	900.9	379	6.09	.1643
56.3	71	303.7	274.4	1174.6	900.2	374	6.01	.1665
57.3	72	304.6	275.3	1174.8	899.5	369	5.93	1687
58.3	73	305.6	276.3	1175.1	898.9	365	5.85	.1709
59.3	74	306.5	277.2	1175.4	898.2	360	5.78	.1731
60.3	75	307.4	278.2	1175.7	897.5	356	5.71	.1753
61.3	76	308.3	279.1	1176.0	896.9	351	5.63	.1775
62.3	77	309.2	280.0	1176.2	896.2	347	5.57	.1797
63.3	78	310.1	280.9	1176.5	895.6	343	5.50	.1819
64.3	79	310.9	281.8	1176.8	895.0	339	5.43	.1840
65.3	80	311.8	282.7	1177.0	894.3	334	5.37	.1862
66.3	81	312.7	283.6	1177.3	893.7	331	5.31	.1884
67.3	82	313.5	284.5	1177.6	893.1	327	5.25	.1906
68.3	83	314.4	285.3	1177.8	892.5	323	5.18	.1928
69.3	84	315.2	286.2	1178.1	891.9	320	5.13	.1950
70.3	85	316.0	287.0	1178.3	891.3	316	5.07	.1971
71.3	86	316.8	287.9	1178.6	890.7	313	5.02	.1993
72.3	87	317.7	288.7	1178.8	890.1	309	4.96	.2015
73.3	88	318.5	289.5	1179.1	889.5	306	4.91	.2036
74.3	89	319.3	290.4	1179.3	888.9	303	4.86	.2058
75.3	90	320.0	291.2	1179.6	888.4	299	4.81	.2080

TABLE 4—*Continued*

Gauge Pressure Lbs. per Sq. In.	Absolute Pressure Lbs. per Sq. In.	Temp. Degrees F.	Total Heat above 32° F. In the Water h Heat-units	Total Heat above 32° F. In the Steam H Heat-units	Latent Heat H-h Heat-units	Relative Volume	Cubic Feet in 1 Lb. Wt. of Steam	Wt. of 1 Cubic Foot of Steam, Lbs.
76.3	91	320.8	292.0	1179.8	887.8	296	4.76	.2102
77.3	92	321.6	292.8	1180.0	887.2	293	4.71	.2123
78.3	93	322.4	293.6	1180.3	886.7	290	4.66	.2145
79.3	94	323.1	294.4	1180.5	886.1	287	4.62	.2166
80.3	95	323.9	295.1	1180.7	885.6	285	4.57	.2188
81.3	96	324.6	295.9	1181.0	885.0	282	4.53	.2210
82.3	97	325.4	296.7	1181.2	884.5	279	4.48	.2231
83.3	98	326.1	297.4	1181.4	884.0	276	4.44	.2253
84.3	99	326.8	298.2	1181.6	883.4	274	4.40	.2274
85.3	100	327.6	298.9	1181.8	882.9	271	4.36	.2296
86.3	101	328.3	299.7	1182.1	882.4	268	4.32	.2317
87.3	102	329.0	300.4	1182.3	881.9	266	4.28	.2339
88.3	103	329.7	301.1	1182.5	881.4	264	4.24	.2360
89.3	104	330.4	301.9	1182.7	880.8	261	4.20	.2382
90.3	105	331.1	302.6	1182.9	880.3	259	4.16	.2403
91.3	106	331.8	303.3	1183.1	879.8	257	4.12	.2425
92.3	107	332.5	304.0	1183.4	879.3	254	4.09	.2446
93.3	108	333.2	304.7	1183.6	878.8	252	4.05	.2467
94.3	109	333.9	305.4	1183.8	878.3	250	4.02	.2489
95.3	110	334.5	306.1	1184.0	877.9	248	3.98	.2510
96.3	111	335.2	306.8	1184.2	877.4	246	3.95	.2531
97.3	112	335.9	307.5	1184.4	876.9	244	3.92	.2553
98.3	113	336.5	308.2	1184.6	876.4	242	3.88	.2574
99.3	114	337.2	308.8	1184.8	875.9	240	3.85	.2596
100.3	115	337.8	309.5	1185.0	875.5	238	3.82	.2617
101.3	116	338.5	310.2	1185.2	875.0	236	3.79	.2638
102.3	117	339.1	310.8	1185.4	874.5	234	3.76	.2660
103.3	118	339.7	311.5	1185.6	874.1	232	3.73	.2681
104.3	119	340.4	312.1	1185.8	873.6	230	3.70	.2703
105.3	120	341.0	312.8	1185.9	873.2	228	3.67	.2764
106.3	121	341.6	313.4	1186.1	872.7	227	3.64	.2745
107.3	122	342.2	314.1	1186.3	872.3	225	3.62	.2766
108.3	123	342.9	314.7	1186.5	871.8	223	3.59	.2788
109.3	124	343.5	315.3	1186.7	871.4	221	3.56	.2809
110.3	125	344.1	316.0	1186.9	870.9	220	3.53	.2830
111.3	126	344.7	316.6	1187.1	870.5	218	3.51	2851
112.3	127	345.3	317.2	1187.3	870.0	216	3.48	.2872
113.3	128	345.9	317.8	1187.4	869.6	215	3.46	.2894

TABLE 4—*Continued*

Gauge Pressure Lbs. per Sq. In.	Absolute Pressure Lbs. per Sq. In.	Temp. Degrees F.	Total Heat Above 32° F.		Latent Heat H-h Heat-units	Relative Volume	Cubic Feet in 1 Lb. Wt. of Steam	Wt. of 1 Cubic Foot of Steam, Lbs.
			In the Water h Heat-units	In the Steam H Heat-units				
114.3	129	346.5	318.4	1187.6	869.2	213	3.43	.2915
115.3	130	347.1	319.1	1187.8	868.7	212	3.41	.2936
116.3	131	347.6	319.7	1188.0	868.3	210	3.38	.2957
117.3	132	348.2	320.3	1188.2	867.9	209	3.36	.2978
118.3	133	348.8	320.8	1188.3	867.5	207	3.33	.3000
119.3	134	349.4	321.5	1188.5	867.0	206	3.31	.3021
120.3	135	350.0	322.1	1188.7	866.6	204	3.29	.3042
121.3	136	350.5	322.6	1188.9	866.2	203	3.27	.3063
122.3	137	351.1	323.2	1189.0	865.8	201	3.24	.3084
123.3	138	351.8	323.8	1189.2	865.4	200	3.22	.3105
124.3	139	352.2	324.4	1189.4	865.0	199	3.20	.3126
125.3	140	352.8	325.0	1189.5	864.6	197	3.18	.3147
126.3	141	353.3	325.5	1189.7	864.2	196	3.16	.3169
127.3	142	353.9	326.1	1189.9	863.8	195	3.14	.3190
128.3	143	354.4	326.7	1190.0	863.4	193	3.11	.3211
129.3	144	355.0	327.2	1190.2	863.0	192	3.09	.3232
130.3	145	355.5	327.8	1190.4	862.6	191	3.07	.3253
131.3	146	356.0	328.4	1190.5	862.2	190	3.05	.3274
133.3	148	357.1	329.5	1190.9	861.4	187	3.02	.3316
135.3	150	358.2	330.6	1191.2	860.6	185	2.98	.3358
140.3	155	360.7	333.2	1192.0	858.7	179	2.89	.3463
145.3	160	363.3	335.9	1192.7	856.9	174	2.80	.3567
150.3	165	365.7	338.4	1193.5	855.1	169	2.72	.3671
155.3	170	368.2	340.9	1194.2	853.3	164	2.65	.3775
160.3	175	370.5	343.4	1194.9	851.6	160	2.58	.3879
165.3	180	372.8	345.8	1195.7	849.9	156	2.51	.3983
170.3	185	375.1	348.1	1196.3	848.2	152	2.45	.4087
175.3	190	377.3	350.4	1197.0	846.6	148	2.39	.4191
180.3	195	379.5	352.7	1197.7	845.0	144	2.33	.4296
185.3	200	381.6	354.9	1198.3	843.4	141	2.27	.4400
190.3	205	383.7	357.1	1199.0	841.9	138	2.22	.4503
195.3	210	385.7	359.2	1199.6	840.4	135	2.17	.4605
200.3	215	387.7	361.3	1200.2	838.9	132	2.12	.4707
205.3	220	389.7	362.2	1200.8	838.6	129	2.06	.4852
245.3	260	404.4	377.4	1205.3	827.9	110	1.76	.5686
285.3	300	417.4	390.9	1209.2	818.3	96	1.53	.6515
485.3	500	467.4	443.5	1224.5	781.0	59	.94	1.062
685.3	700	504.1	482.4	1235.7	753.3	42	.68	1.470
985.3	1000	546.8	528.3	1248.7	720.3	30	.48	2.082

Questions

1. What is one of the most important of a fireman's duties?

2. What should the fireman attend to first of all when getting his engine ready to start out from the roundhouse?

3. What other details should be looked after at this time?

4. What condition should his fire be in before leaving a terminal?

5. What is the proper depth of fire to be carried?

6. Should the fireman read and understand the train orders?

7. How should the coal be supplied to the fire while running?

8. What is the best rule for the fireman to observe?

9. What precautions should a fireman practice respecting admission of air to the fire-box?

10. How may he prevent the grates from being burned out?

11. Explain why the exhaust creates such a strong draft.

12. When should the blower be used?

13. Why is a larger grate area required for hard coal than for soft coal?

14. Describe a water grate.

15. How many square feet of grate surface is needed to burn one ton of soft coal per hour?

16. For what purpose is a steam gauge connected to a boiler?

17. Explain the construction and working of the Bourdon spring gauge.

18. How should steam gauges be tested?

19. For what purpose is a safety valve used?

20. How many pop valves should a locomotive boiler be equipped with?

21. Explain the working of a pop valve.

22. What are the fireman's duties upon arrival at a terminal?

23. What is combustion?

24. What is one of the main factors in combustion?

25. Of what is air composed?

26. In what proportion are these two gases combined?

27. What is the principal constituent of coal and other fuels?

28. What other valuable constituent is contained in bituminous coal?

29. What is the usual temperature of a boiler furnace when in active operation?

30. About what should be the temperature of the escaping gases?

31. What two factors are indispensable in the economical use of coal?

32. What is heat?

33. What is the heat unit?

34. What is the mechanical equivalent of heat?

35. How many heat units are there in one pound of carbon?

36. How many heat units are there in one pound of hydrogen gas?

37. What is specific heat?

38. What is sensible heat?

39. What is latent heat?

40. Is the latent heat imparted to a body lost?

41. What is meant by the total heat of evaporation?

42. How much heat expressed in heat units is re-

quired to evaporate one pound of water from a temperature of 32° into steam at atmospheric pressure?

43. Name the two elements composing pure water.

44. In what proportion are these two gases combined in the formation of water?

45. Is perfectly pure water desirable for use in steam boilers?

46. What causes scale to form in boilers?

47. What proportion of mineral matter is usually found in water?

48. What is steam?

49. Of what does matter consist?

50. How does the application of heat to any substance affect its molecules?

51. In what particular manner does heat affect the molecules of water?

52. Is steam under pressure visible?

53. What is saturated steam?

54. What is dry steam?

55. What is superheated steam?

56. What is meant by the term total heat in steam?

57. What is meant by the density of steam?

58. What is meant by the volume of steam?

59. What is the weight of a cubic foot of water at 39.1° temperature?

60. What is the weight of a cubic foot of water at a temperature of 212°?

61. What is the boiling point of water in the open air at sea level?

62. At what temperature will water boil in a perfect vacuum?

63. What is meant by the relative volume of steam?

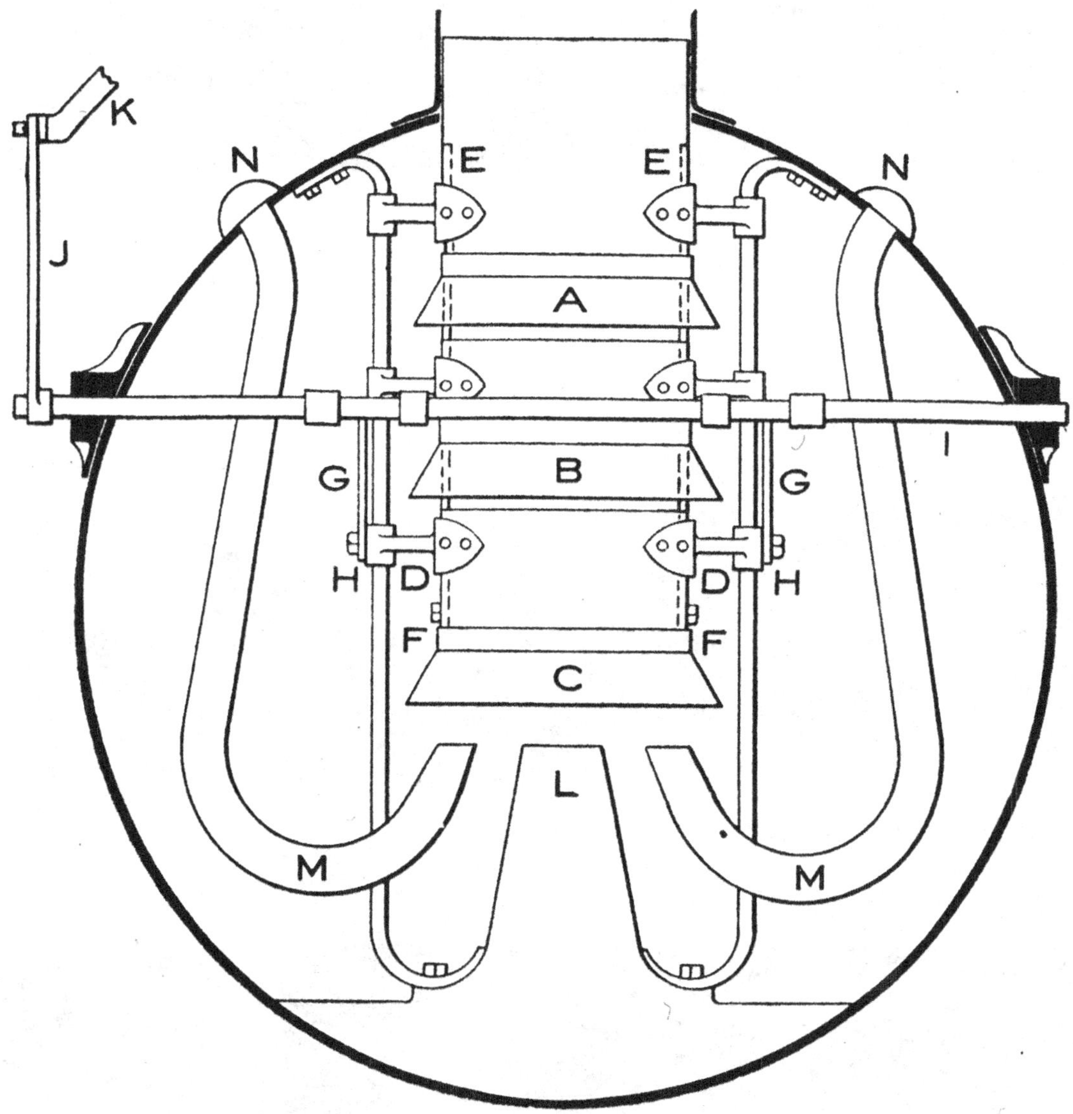
K
N
N
E
E
J
A
B
G
G
I
H
D
D
H
F
F
C
L
M
M

1026

CHAPTER II

THE BOILER

In order that the student may get a general idea of the construction of a locomotive boiler, a sectional elevation of one is shown in Fig. 6.

The four vital organs of a locomotive boiler are: first, the fire-box A; second, the cylinder or barrel B-B; third, the flues or tubes C-C, and fourth, the smokestack D. Underneath the fire-box is suspended the ash pan E, next above the ash pan appears the mud ring F-F. This is a wrought iron bar bent to the proper form to extend around the bottom of the inside of the fire-box, the ends welded, and the ring thus formed is then drilled and riveted to the inside and outside sheets.

The fire-box is a rectangular box constructed of steel plates G-G from 3/8 to 9/16 in. in thickness. The inner shell is surrounded by an outside shell H-H, also constructed of steel plates, usually of about the same thickness as the inner plates. The outside shell is enough larger than the inside one to allow a space of 2½ to 4½ in. between the inner and outer plates. This space is called the water space and entirely surrounds the fire-box on the four sides, the water occupying it being in free communication with the main body of water in the boiler It will thus be seen that the flat sides of the fire-box are subjected to the full pressure of the steam, and unless they be strengthened in some manner they will bulge apart. This danger is obviated by the use of stay bolts J-J, Fig. 6. These are made of the best quality of wrought iron, generally from ⅞ to $1\frac{1}{16}$ in. in diameter, and have a screw thread cut their whole length. They are screwed through both the outside and inside plates at intervals of from 4 to 4½ in. apart center to center, thus securely binding the plates together. The projecting ends of these stay bolts are also riveted down onto the plates, thus further increasing their holding power.

Owing to the unequal expansion and contraction of the inner and outer plates, stay bolts are subjected to great strains and very frequently break, thereby causing a large amount of trouble. They should be made

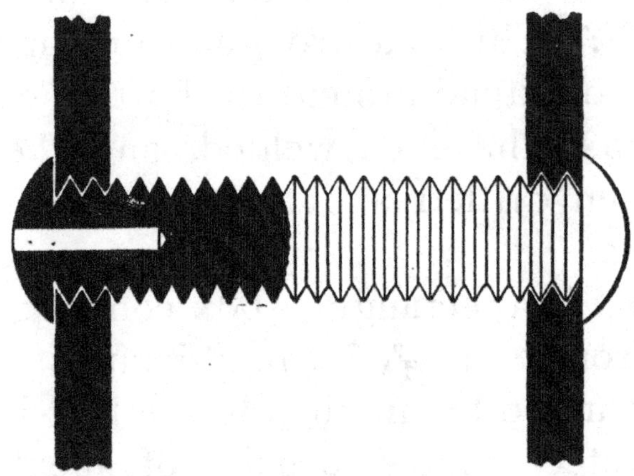

FIGURE 7

tubular, or at least have a small hole drilled into one end, as shown in Fig. 7, extending into the bolt a distance greater than the thickness of the outside plate, so that if the bolt breaks which generally occurs next the outside plate, the water will escape through the fracture into the hole and thus indicate the defect and the danger.

The Tate flexible stay bolt, which received the highest award at the St. Louis exposition in 1904, appears to offer at least a partial solution of the problem of staying fire-box sheets. Fig. 8 is a sectional view showing the design of this stay bolt. The ball-shaped head of the bolt C is inclosed within a socket formed by a sleeve B that screws into the outer sheet, and a cap A that screws onto the sleeve. The other end of the bolt is screwed into and through the fire sheet a

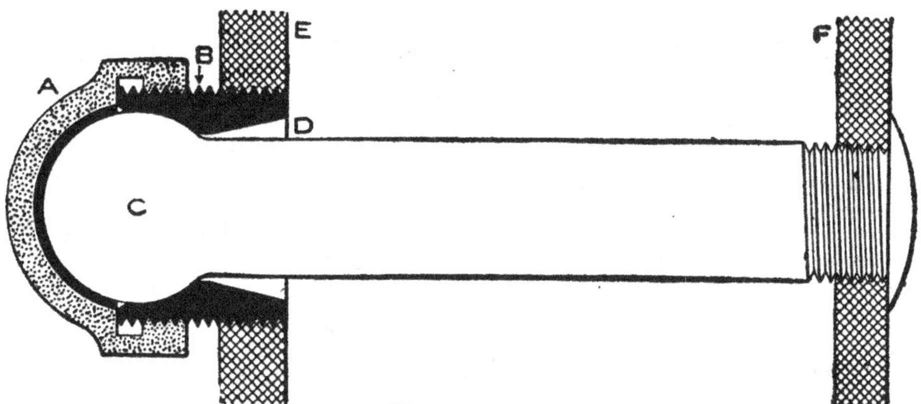

FIGURE 8

sufficient distance to allow of riveting. It is apparent that the freedom of movement of the head of the bolt within its socket will allow the fire sheet to go and come, without subjecting the bolt to such severe strains and transverse stresses as would occur if the bolt were rigid. Fig. 9 is a full view of the bolt,

FIGURE 9

except that the thread has not yet been cut on the end that screws into the fire sheet. The Tate flexible stay bolt is manufactured by the Flannery Bolt Company, Pittsburg, Pa.

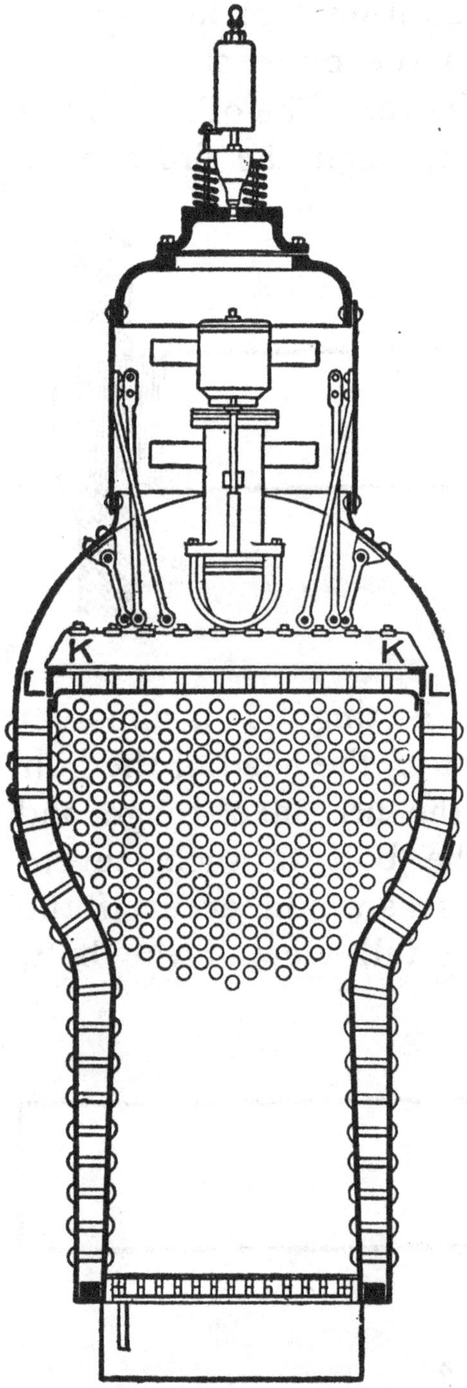

FIGURE 10

It is also necessary to strengthen the flat top or crown sheet of the fire-box. There are three common methods by which this is done: first, by crown bars; second, by radial stays, and third, by the Belpaire system.

In Fig. 6 the crown bar method is shown, K-K being the ends of the crown bars Fig. 10 is a transverse sectional view of the same boiler, and one of the crown bars, K-K, is shown extending across the top of the fire-box above the crown sheet and supported at the ends by special castings that rest on the edges of the side sheets and on the flange of the crown sheet at L-L. These crown bars are double girders, and a space is allowed between them and the top of the crown sheet to allow the water to circulate freely. At intervals of 4 or 5 in. crown bolts are placed having the head inside the fire-box and the nut bearing on a plate on top of the crown bar. There is also a thimble or ring for each bolt to pass through, between the top of the crown sheet and the bottom of the crown bars. These thimbles maintain the proper distance between the crown sheet and crown bars.

The second method of supporting the crown sheet is by the use of radial stays, which are long stay bolts screwed into the outer shell and into the crown sheet.

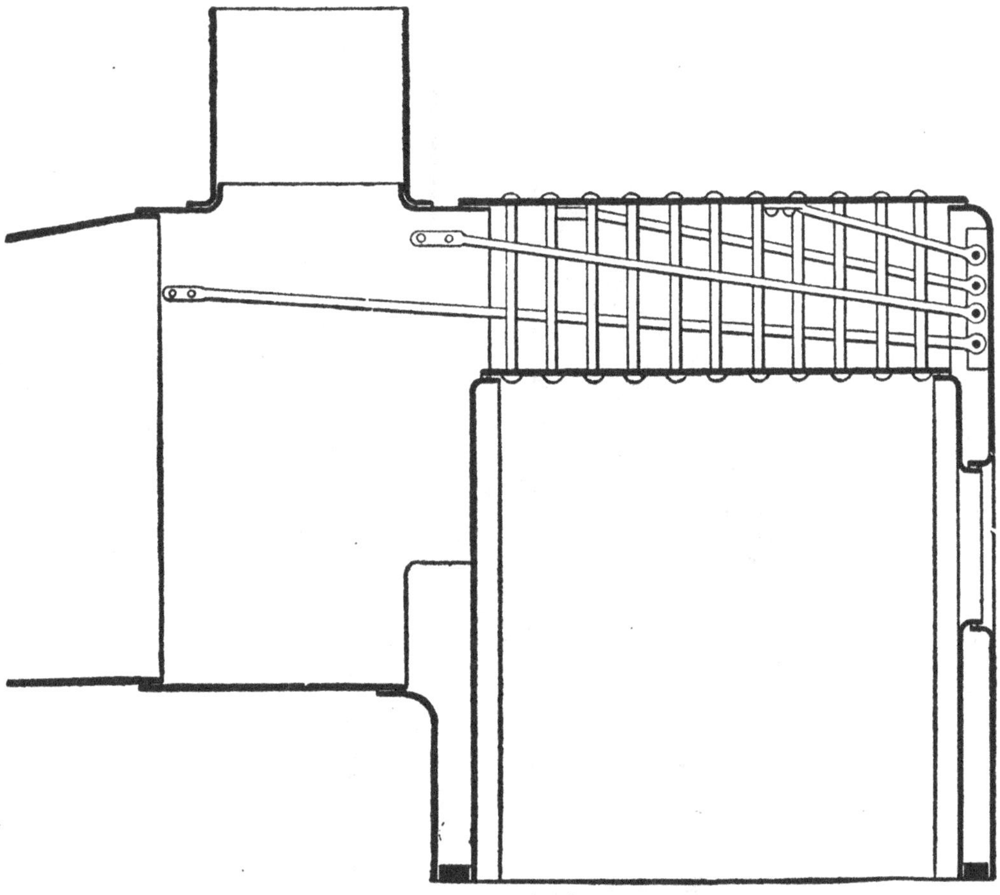

FIGURE 11

Fig. 11 shows a longitudinal section of a fire-box having the crown sheet secured by radial stays, and Fig. 12 is a transverse section and back view of the same. The principal defect in this construction is, that in order to resist successfully the strains induced by the pressure on the crown sheet, the stays should be placed at right angles to its surface, and in order to resist the pressure on the outer shell they should be radial to its cylindrical form, but as it is impossible to so locate them the strains are not equally divided and a certain distortion of both the stays and the sheets is the result. The only thing that can be done under

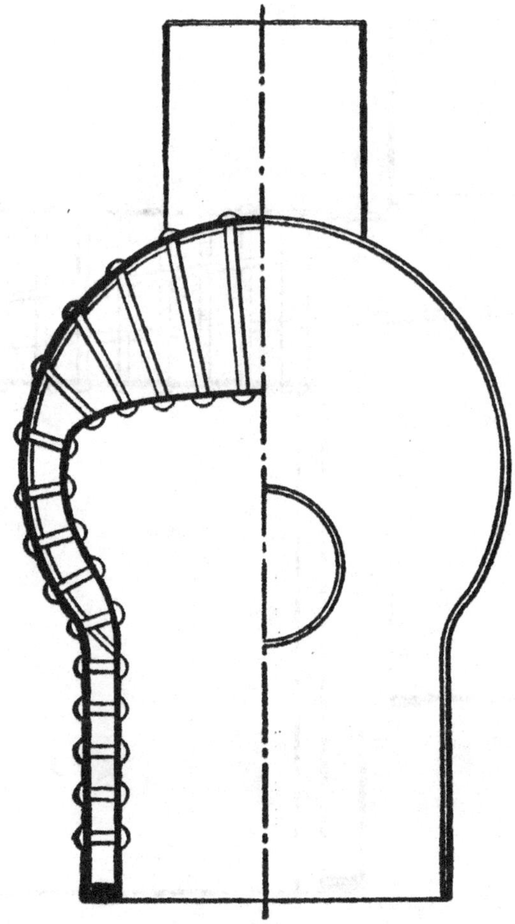

FIGURE 12

such conditions is to approximate as closely as possible the correct position of the stays

In the third or Belpaire system the outside shell of the boiler directly over the crown sheet is made flat to conform to the surface of the crown sheet. This permits of positive staying, the stays all having good bearings in and on the sheets. This method is illustrated by Figs. 13 and 14, which show longitudinal and transverse sections of this form of fire-box. The long stays S S S are seen to be connected at right angles to the flat plates, and the sides, which are also flat, are braced by the rods B B B extending across from side to side. A great advantage in this form of fire-box is

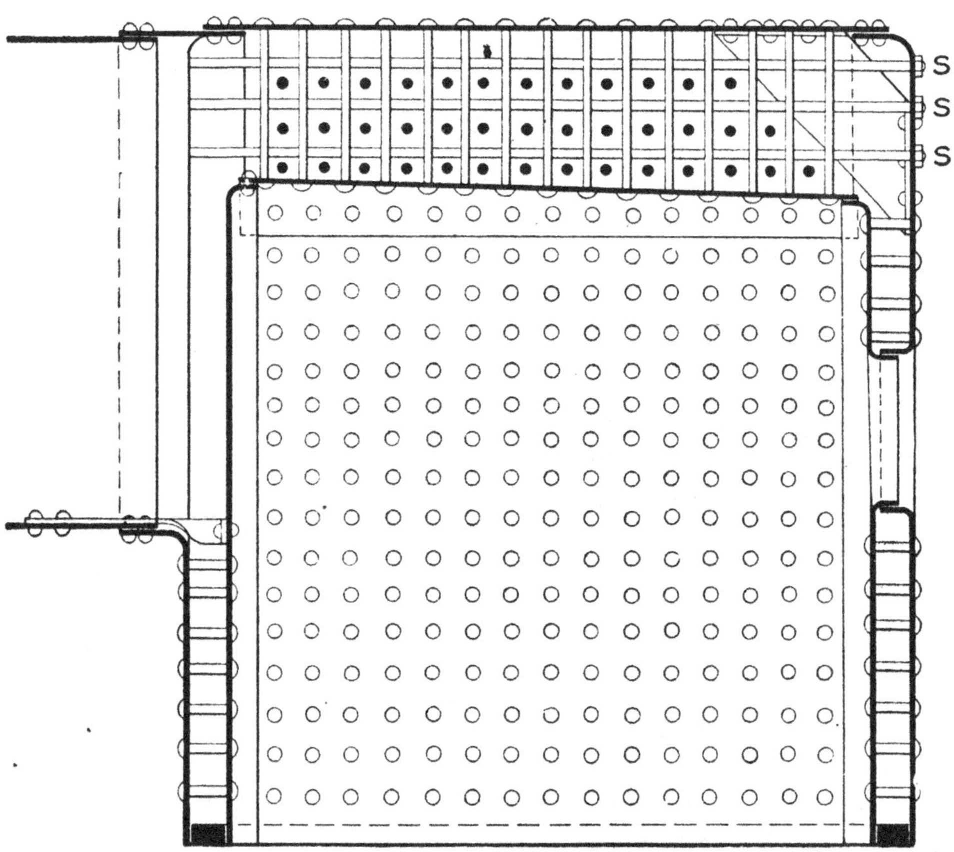

FIGURE 13

that the crown sheet and the flat outside sheet directly over it have more or less flexibility and are free to bend or spring, according as the inside plates become heated and expand, or cool and contract. On the other hand, if the outside sheet is cylindrical in shape and has the crown sheet stayed to it by means of radial stays, it will be subjected to excessive distortional strains caused by the more or less pushing upwards of the stays as the inner plates become heated.

The crown sheets of locomotive boilers are as a rule made to slope downwards from the front end of the fire-box toward the back end, so as to be several inches lower behind than in front. This is done in order to lessen the danger of the back end of the

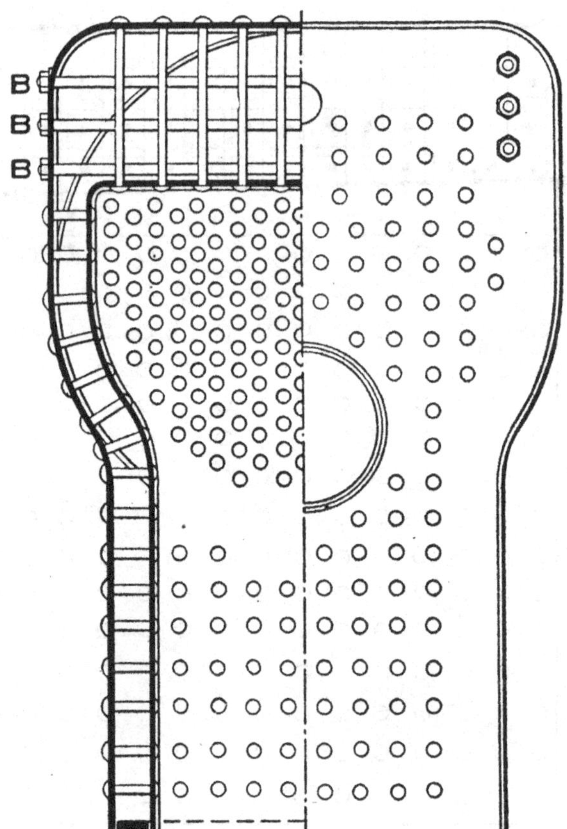

FIGURE 14

crown sheet becoming uncovered of water in running down a steep grade. There is not so much danger of the front end of the crown sheet becoming uncovered, either in going up or down a grade, for the reason that it is nearer the center of the length of the boiler.

The usual method of staying the heads of locomotive boilers is illustrated in Fig. 6. Diagonal stays or braces S S S S are used, having one end riveted to the shell and the other end connected to that portion of the head that needs bracing.

The flues serve to brace the flue sheet and all of that portion of the front head to which they are connected. Sometimes gusset stays are used for staying the heads. A gusset stay is a triangular piece of boiler plate P, Fig. 15, connected to the boiler head H and to the shell S by means of angle irons A A A A, which are riveted to the head. The plate P is connected to the angle irons by rivets. The tube plates or flue sheets

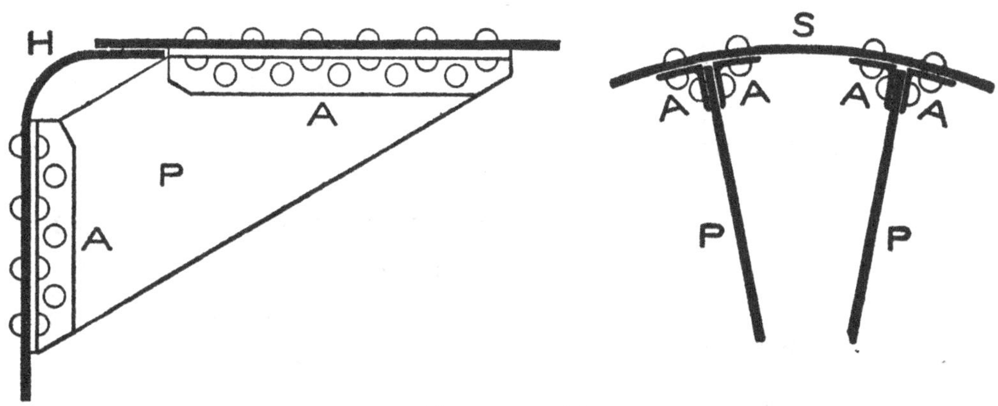

FIGURE 15

are of necessity thicker than the shell, owing to the fact that they are considerably weakened by the holes drilled in them for the tubes. By reference to Fig. 6 the arrangement of the tubes will be clearly understood, N being the fire-box end and M the smoke-box end. Fig. 10 gives a view of the fire-box end of the tubes, which in this case are arranged in vertical rows. In some cases the tubes are placed in horizontal rows. Opinions differ as to the best arrangement, but it is generally conceded that the plan of having them in vertical rows permits of a freer circulation of the water around them.

The diameter of locomotive tubes is usually two inches, as that size has been found by experience to be the most suitable for the distribution of the hot gases on their way from the fire-box to the smoke-stack.

The tubes or flues are made water-tight in the sheets by being expanded in the holes drilled to receive them. The ends of the tubes are allowed to project through the sheets ¼ in. or more. Copper ferrules are generally slipped in over the outside of the tubes, and the tube is then expanded to fill the hole and a water-tight joint is thus secured. After the tube has been sufficiently expanded, the projecting end is turned back onto the sheet and formed into a bead by the use of a caulking tool made especially for the purpose. Fig. 16 is a sectional view of one end of a tube as it appears after being expanded into the sheet.

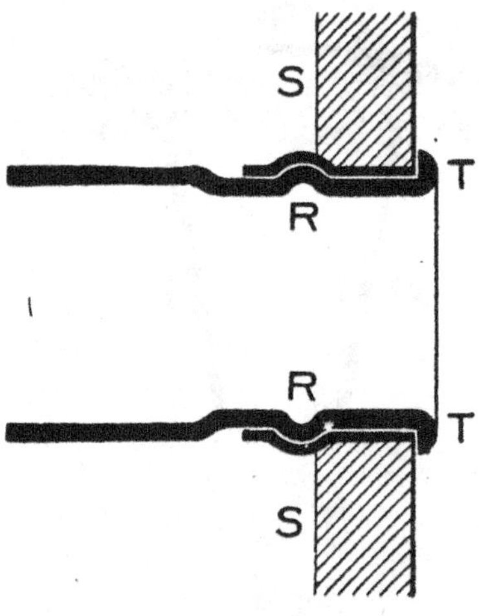

FIGURE 16

There have been various types of tools designed and made for expanding tubes, but the two most generally used are the Prosser, Figs 17 and 18, and the Dudgeon, Fig 19.

The Prosser tube expander is an expanding plug made up of eight or more sectors, 1, 2, 3, 4, 5, 6, 7, 8, held together by an open steel ring or spring clasp C (see Fig. 18). The sector-shaped pieces have their inner edges cut away in such shape as to leave a tapered hole H through the center of the plug. Into this hole the tapered mandrel E is inserted, and when the expander is inserted into the mouth of the tube and the mandrel driven in, the sectors will be slightly separated and the tendency will be to expand the tubes. The outside conformation of the sectors composing the plug is such that, when the tube is expanded, it not only completely fills the hole in the

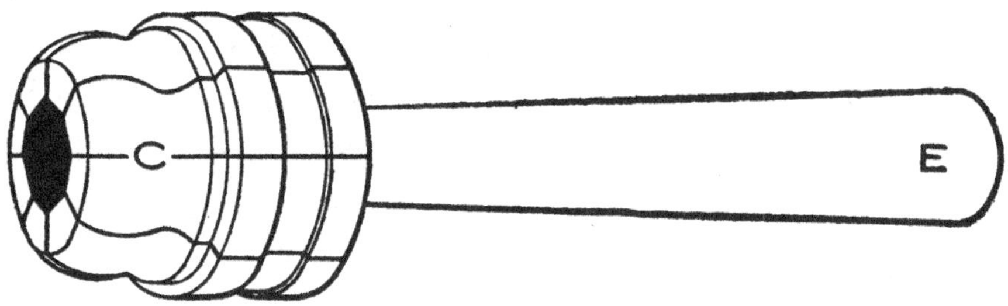

FIGURE 17

tube sheet but is also expanded past the edge of the hole, both on the inside and outside of the sheet, thus securely binding the tube in the sheet and causing it to act as a brace. Referring to Fig. 16, S S is the tube sheet, R R shows the expanded ridge on the tube inside the sheet, and T T indicates the manner in which the end of the tube is expanded and beaded over onto the outer edge of the hole.

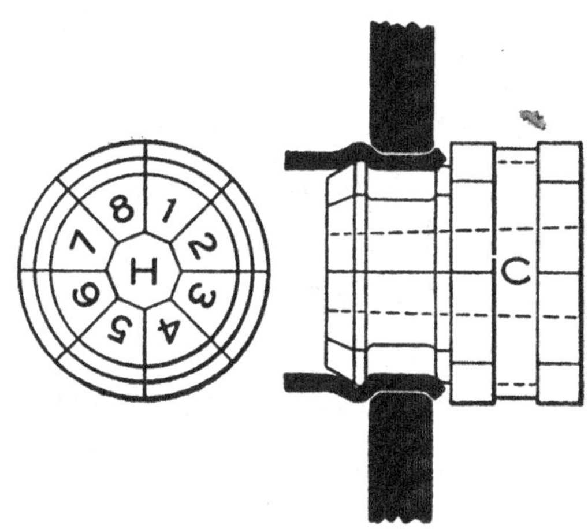

FIGURE 18

The Dudgeon roller tube expander, shown in Fig. 19, consists of a hollow plug having a sleeve or cap at one end that bears against the outside of the sheet, thus serving as a guide to the roller when in use. Three cavities are cut longitudinally in the plug, and into each one of these cavities a roller is inserted which is free to revolve. These rollers can also move a short distance outward from the center of the plug. In using this expander the plug is inserted into the mouth of the tube as far as the cap will permit. A tapered mandrel is then driven into the central opening, and the rollers are forced out against the inner surface of the tube. The mandrel is then slowly turned around by means of a short steel rod inserted into one of the holes shown in the head (see Fig. 19). This causes the plug to revolve, as well as the rollers which bear hard against the tube, and expand it so as to fill the hole in the sheet.

The Dudgeon expander is also a very efficient tool for repairing leaky tubes. Cast iron or steel ferrules

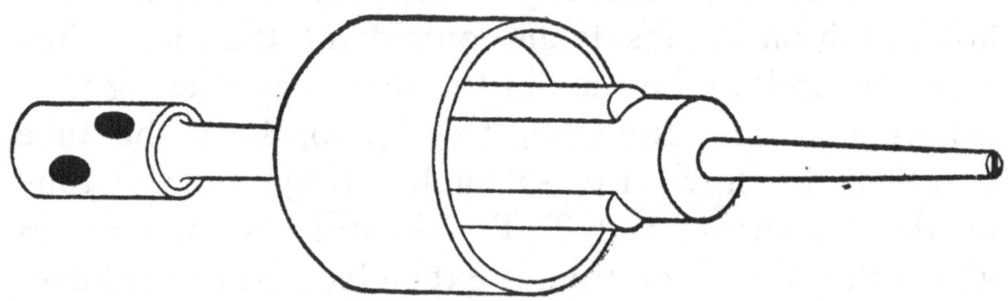

FIGURE 19

made slightly tapering are sometimes driven into the mouths of tubes after they have been expanded, but this method, although it may serve to prevent leakage, will at the same time decrease the capacity of the tubes to conduct the heat.

As the term tensile strength (T. S.) will be used quite frequently in the remaining portion of this chapter, it is proper that its meaning be explained for the benefit of the beginner.

The expression tensile strength per square inch as referring to a boiler sheet means that when the plate is rolled, and before it is accepted by the inspector, a small test piece having a sectional area of one square inch is cut from the plate and placed in a testing machine, where it is subjected to a pull or strain in the direction of its length, and this strain must equal the T. S. called for in the specifications. If the specifications call for a T. S. of 66,000 lbs. per square inch, the test piece must withstand that much of a strain before showing signs of breaking, otherwise the sheet will or should be rejected.

When steel was first introduced as a material for boiler plate, it was customary to demand a high tensile strength, 70,000 to 74,000 lbs. per square inch, but experience and practice demonstrated in course of time that it was much safer to use a material of lower tensile strength. It was found that with steel boiler plate of high tenacity there was great liability of its cracking, and also of certain changes occurring in its physical properties, brought about by the variations in temperature to which it was exposed. Consequently present-day specifications for steel boiler plate call for tensile strengths running from 55,000 to 66,000 lbs., usually 60,000 lbs. per square inch. Dr. Thurston gives what he calls "good specifications" for boiler steel as follow: "Sheets to be of uniform thickness, smooth finish, and sheared closely to size ordered. Tensile strength to be 60,000 lbs. per square inch for fire-box sheets and 55,000 lbs. per square inch for shell sheets. Working test: a piece from each sheet to be heated to a dark cherry red, plunged into water at 60° and bent double, cold, under the hammer. Such piece to show no flaw after bending." The U. S. Board of Supervising Inspectors of Steam Vessels prescribes, in Section 3 of General Rules and Regulations, the following method for ascertaining the tensile strength of steel plate for boilers: "There shall be taken from each sheet to be used in shell or other parts of boiler which are subject to tensile strain, a test piece prepared in form according to the following diagram:

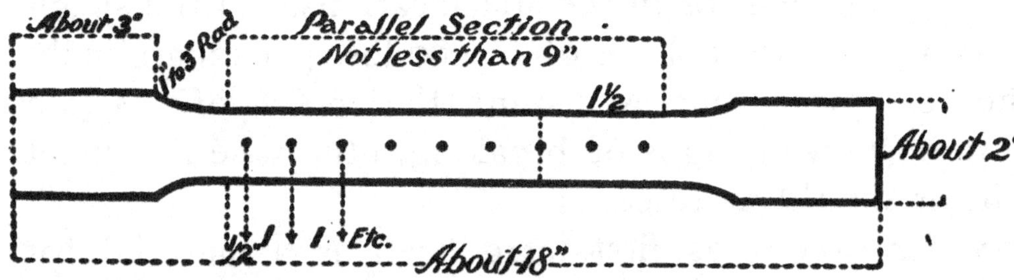

TEST PIECE

The straight part in center shall be 9 in. in length and 1 in. in width, marked with light prick punch marks at distances 1 in. apart, as shown, spaced so as to give 8 in. in length. The sample must show, when tested, an elongation of at least 25 per cent in a length of 2 in. for thickness up to ¼ in. inclusive; in a length of 4 in., for over ¼ in. to $\frac{7}{16}$ in. inclusive; in a length of 6 in., for all plates over $\frac{7}{16}$ in. and under 1¾ in. in thickness. The samples shall also be capable of being bent to a curve of which the inner radius is not greater than 1½ times the thickness of the plates, after having been heated uniformly to a low cherry red and quenched in water of 82° F."

Punched and Drilled Plates. Much has been written on this subject, and it is still open for discussion. If the material is a good, soft steel, punched sheets are apparently as strong and in some instances stronger than drilled; especially is this the case with regard to the shearing resistance of the rivets, which is greater with punched than with drilled holes.

Concerning rivets and rivet iron and steel Dr. Thurston has this to say in his "Manual of Steam

Boilers": "Rivet iron should have a tenacity in the bar approaching 60,000 lbs. per square inch, and should be as ductile as the very best boiler plate when cold. A good 5/8-in. iron rivet can be doubled up and hammered together cold without exhibiting a trace of fracture." The shearing resistance of iron rivets is about 85 per cent and that of steel rivets about 77 per cent of the tenacity of the original bar, as shown by experiments made by Greig and Eyth. The researches made by Wöhler demonstrated that the shearing strength of iron was about four-fifths of the tensile strength.

The tables that follow have been compiled from the highest authorities and show the results of a long and exhaustive series of tests and experiments made in order to ascertain the proportions of riveted joints that will give the highest efficiencies.

The following table gives the diameters of rivets for various thicknesses of plates and is calculated according to a rule given by Unwin.

Table 5

Table of Diameters of Rivets*

Thickness of Plate	Diameter of Rivet	Thickness of Plate	Diameter of Rivet
1/4 inch	1/2 inch	9/16 inch	7/8 inch
5/16 “	9/16 “	5/8 “	15/16 “
3/8 “	11/16 “	3/4 “	1 1/16 “
7/16 “	3/4 “	7/8 “	1 1/8 “
1/2 “	13/16 “	1 “	1 1/4 “

The efficiency of the joint is the percentage of the strength of the solid plate that is retained in the joint,

*Machine design—W. C. Unwin.

and it depends upon the kind of joint and method of construction.

If the thickness of the plate is more than ½ in., the joint should always be of the double butt type.

The diameters of rivets, rivet holes, pitch and efficiency of joint, as given in the following table, which was published in the "Locomotive" several years ago, were adopted at the time by some of the best establishments in the United States.*

Table 6

Proportions and Efficiencies of Riveted Joints

	Inch	Inch	Inch	Inch	Inch
Thickness of plate	1/4	5/16	3/8	7/16	1/2
Diameter of rivet	5/8	11/16	3/4	13/16	7/8
Diameter of rivet-hole	11/16	3/4	13/16	7/8	15/16
Pitch for single riveting	2	2 1/16	2 1/8	2 3/16	2 1/4
Pitch for double riveting	3	3 1/8	3 1/4	3 3/8	3 1/2
Efficiency—single-riveted joint	.66	.64	.62	.60	.58
Efficiency—double-riveted joint	.77	.76	.75	.74	.73

Concerning the proportions of double-riveted butt joints, Professor Kent says: "Practically it may be said that we get a double-riveted butt joint of maximum strength by making the diameter of the rivet about 1.8 times the thickness of the plate, and making the pitch 4.1 times the diameter of the hole."

Table 7, as given below, is condensed from the report of a test of double-riveted lap and butt joints.† In this test the tensile strength of the plates was 56,000 to

*Thurston's Manual of Steam Boilers.

† Proc. Inst. M. E., Oct., 1888.

58,000 lbs. per square inch, and the shearing resistance of the rivets (steel) was about 50,000 lbs. per square inch.

TABLE 7

DIAMETER AND PITCH OF RIVETS—DOUBLE-RIVETED JOINT

Kind of Joint	Thickness of Plate	Diameter of Rivet	Ratio of Pitch to Diameter
Lap	3/8 inch	0.8 inches	3.6 inches
Butt	3/8 "	0.7 "	3.9 "
Butt	3/4 "	1.1 "	4.0 "
Butt	1 "	1.3 "	3.9 "

Lloyd's rules, condensed, are as follows:

LLOYD'S RULES—THICKNESS OF PLATE AND DIAMETER OF RIVETS

Thickness of Plate	Diameter of Rivets	Thickness of Plate	Diameter of Rivets
3/8 inch	5/8 inch	3/4 "	7/8 inch
7/16 "	5/8 "	13/16 "	7/8 "
1/2 "	3/4 "	7/8 "	1 "
9/16 "	3/4 "	15/16 "	1 "
5/8 "	3/4 "	1 "	1 "
11/16 "	7/8 "		

The following Table 8 is condensed from one calculated by Professor Kent,* in which he assumes the shearing strength of the rivets to be four-fifths of the tensile strength of the plate per square inch, and the excess strength of the perforated plate to be 10 per cent.

*Kent's Mechanical Engineer's Pocket-Book, page 362.

TABLE 8

Thickness of Plate	Diameter of Hole	Pitch		Efficiency	
		Single Riveting	Double Riveting	Single Riveting	Double Riveting
Inches	Inches	Inches	Inches	Per Cent	Per Cent
3/8	7/8	2.04	3.20	57.1	72.7
7/16	1	2.30	3.61	56.6	72.3
1/2	1	2.14	3.28	53.3	70.0
1/2	1 1/8	2.57	4.01	56.2	72.0
9/16	1	2.01	3.03	50.4	67.0
9/16	1 1/8	2.41	3.69	53.3	69.5
9/16	1 1/4	2.83	4.42	55.9	71.5
5/8	1	1.91	2.82	47.7	64.6
5/8	1 1/8	2.28	3.43	50.7	67.3
5/8	1 1/4	2.67	4.10	53.3	69.5

Another table of joint efficiences as given by Dr. Thurston* is as follows, slightly condensed from the original calculation:

TABLE 9

Single riveting

Plate thickness.	5/16''	3/8''	7/16''	1/2''	5/8''	3/4''	7/8''	1''
Efficiency	.55	.55	.53	.52	.48	.47	.45	.43

Double riveting

Plate thickness.	3/8''	7/16''	1/2''	3/4''	7/8''	1''
Efficiency......	.73	.72	.71	.66	.64	.63

The author has been at considerable pains to compile Tables 10, 11 and 12, giving proportions and efficiencies of single lap, double lap and butt, and triple-riveted butt joints. The highest authorities have been consulted in the computation of these tables and great care exercised in the calculations.

* Thurston's Manual of Steam Boilers, page 119.

Table 10

Proportions of Single-riveted Lap Joints

Thickness of Plate Inches	Diameter of Rivet Inches	Pitch of Rivet Inches	Efficiency Per Cent
5/16	9/16	1.13	50.5
"	5/8	1.33	53.3
"	11/16	1.55	55.7
3/8	3/4	1.60	53.3
"	7/8	2.04	57.1
7/16	7/8	1.87	53.2
"	1	2.30	56.6
1/2	1	2.14	53.3
"	1 1/8	2.57	56.2
9/16	1	2.01	50.4
"	1 1/8	2.41	53.3
"	1 1/4	2.83	55.9
5/8	1 1/8	2.28	50.7
"	1 1/4	2.67	53.3

It will be noticed that in single-riveted lap joints the highest efficiencies are attained when the diameter of the rivet hole is about 2⅓ times the thickness of the plate, and the pitch of the rivet 2⅜ times the diameter of the hole.

With the double-riveted joint it appears, according to Table 11, that in order to obtain the highest efficiency the joint should be designed so that the diameter of the rivet hole will be from 1⅘ to 2 times the thickness of plate, and the pitch should be from 3⅓ to 3½ times the diameter of the hole. Concerning the thickness of plates Dr. Thurston has this to say:* "Very thin plates cannot be well caulked, and thick plates cannot be safely riveted. The limits are about ¼ of an inch for the lower limit, and ¾ of an inch for the higher limit." The riveting machine, however, overcomes the difficulty with very thick plates.

*Thurston's Manual of Steam Boilers, page 120.

TABLE 11

PROPORTIONS OF DOUBLE-RIVETED LAP AND BUTT JOINTS

Thickness of Plate	Diameter of Rivet	Pitch of Rivet	Efficiency
5/16 inch	9/16 inch	1.71 inches	67.1 per cent
5/16 "	5/8 "	2.05 "	69.5 "
3/8 "	3/4 "	2.46 "	69.5 "
3/8 "	7/8 "	3.20 "	72.7 "
7/16 "	3/4 "	2.21 "	66.2 "
7/16 "	7/8 "	2.86 "	69.4 "
7/16 "	1 "	3.61 "	72.3 "
1/2 "	1 "	3.28 "	70.0 "
1/2 "	1 1/8 "	4.01 "	72.0 "
9/16 "	1 "	3.03 "	67.0 "
9/16 "	1 1/8 "	3.69 "	69.5 "
9/16 "	1 1/4 "	4.42 "	71.5 "
5/8 "	1 1/8 "	3.43 "	67.3 "
5/8 "	1 1/4 "	4.10 "	69.5 "
3/4 "	1 "	2.50 "	72.0 "
7/8 "	1 1/8 "	3.94 "	74.2 "
1 "	1 1/4 "	4.10 "	76.1 "

The triple-riveted butt joint with two welts, one inside and one outside, has two rows of rivets in double shear and one outer row in single shear on each side of the butt, the pitch of rivets in the outer rows being twice the pitch of the inner rows. One of the welts is wide enough for the three rows of rivets each side of the butt, while the other welt takes in only the two close pitch rows.

When properly designed, this form of joint has a high efficiency, and is to be relied upon. Table 12 gives proportions and efficiencies, and it will be noted that the highest degree of efficiency is shown when the diameter of rivet hole is from 1¼ to 1½ times the thickness of plate, and the pitch of the rivets is from 3½ to 4 times the diameter of the hole. This, of course, refers to the pitch of the close rows of rivets, and not the two outer rows.

TABLE 12

PROPORTIONS OF TRIPLE-RIVETED BUTT JOINTS WITH INSIDE AND OUTSIDE WELT

Thickness of Plate Inches	Diameter of Rivet Inches	Pitch of Rivet Inches	Pitch of Outer Rows Inches	Efficiency Per Cent
3/8	13/16	3.25	6.5	84
7/16	13/16	3.25	6.5	85
1/2	13/16	3.25	6.5	83
9/16	7/8	3.50	7.0	84
5/8	1	3.50	7.0	86
3/4	1 1/16	3.50	7.0	85
7/8	1 1/8	3.75	7.5	86
1	1 1/4	3.87	7.7	84

A few examples of calculations for efficiency will be given, taking the three forms of riveted joints in most common use. The following notation will be used throughout:

T.S. = Tensile strength of plate per square inch.
T = Thickness of plate.
C = Crushing resistance of plate and rivets.
A = Sectional area of rivets.
S = Shearing strength of rivets.
D = Diameter of hole (also diameter of rivets when driven).
P = Pitch of rivets.

In the calculations that follow T.S. will be assumed to be 60,000 lbs., S will be taken at 45,000 lbs., and the value of C may be assumed to be 90,000 to 95,000.

Fig. 20 shows a double-riveted lap joint. The style of riveting in this joint is what is known as chain riveting.

In case the rivets are staggered the same rules for calculating the efficiency will hold as with chain riveting, for the reason that with either style of riveting the unit strip of plate has a width equal to the pitch or distance p, Fig. 20.

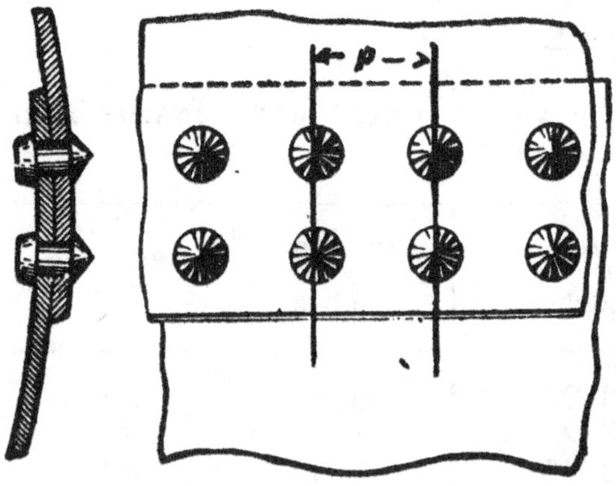

FIGURE 20

The dimensions of the joint under consideration are as follows: P = 3¼ in., T = $\frac{7}{16}$ in., D = 1 in. (which is also diameter of driven rivet).

The strength of the unit strip of solid plate is P × T × T.S. = 85,312.

The strength of net section of plate after drilling is P – D × T × T.S. = 59,062.

The shearing resistance of two rivets is 2A × S = 70,686.

The crushing resistance of rivets and plate is D × 2 × T × C = 78,750.

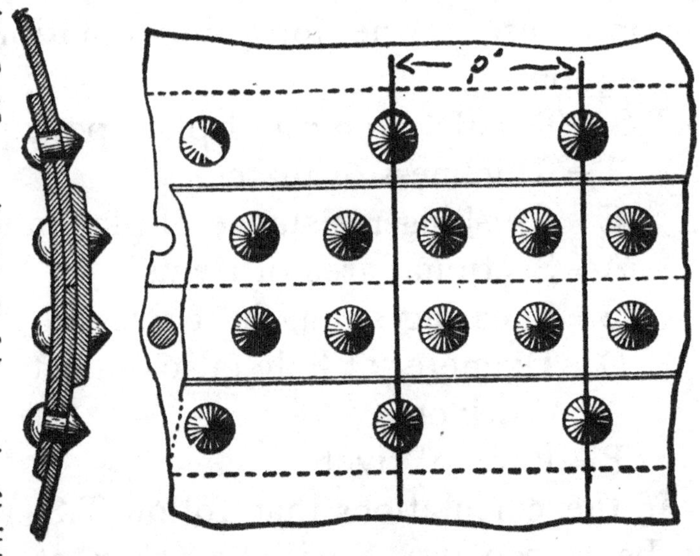

FIGURE 21

It thus appears that the weakest part of the joint is the the net strip or section of plate, the strength of which is 59,062 and the efficiency = 59,062 × 100 ÷ 85,312 = 69.2 per cent.

A double-riveted butt joint is illustrated by Fig. 21, and the dimensions are as follows:

P, inner row of rivets = 2¾ in.

P′, outer row of rivets = 5½ in.

T of plate and butt straps = $\frac{7}{16}$ in.

D of hole and driven rivet = 1 in.

Failure may occur in this joint in five distinct ways, which will be taken up in their order.

1. Tearing of the plate at the outer row of rivets. The net strength at this point is $P - D \times T \times T.S.$, which, expressed in plain figures, results as follows: $5.5 - 1 \times .4375 \times 60{,}000 = 118{,}125$.

2. Shearing two rivets in double shear and one in single shear. Should this occur, the two rivets in the inner row would be sheared on both sides of the plate, thus being in double shear. Opposed to this strain there are four sections of rivets, two for each rivet. Then at the outer row of rivets in the unit strip there is the area of one rivet in single shear to be added. The total resistance, therefore, is $5A \times S$ as follows: $.7854 \times 5 \times 45{,}000 = 176{,}715$.

3. The plate may tear at the inner row of rivets and shear one rivet in the outer row. The resistance in this case would be $P' - 2D \times T \times T.S. + A \times S$ as follows: $5.5 - 2 \times .4375 \times 60{,}000 + .7854 \times 45{,}000 = 127{,}218$.

4. Failure may occur by crushing in front of three rivets. Opposed to this is $3D \times T \times C$, or $1 \times 3 \times .4375 \times 95{,}000 = 124{,}687$.

5. Failure may occur by crushing in front of two rivets and shearing one. The resistance is represented by $2D \times T \times C + 1A \times S$; expressed in figures, $1 \times 2 \times .4375 \times 95{,}000 + .7854 \times 45{,}000 = 118{,}468$.

The strength of a solid strip of plate 5½ in. wide before drilling is $P' \times T \times T.S.$, or $5.5 \times .4375 \times 60{,}000 =$

144,375, and the efficiency of the joint is 118,125 × 100 ÷ 144,375 = 81.1 per cent.

A triple-riveted butt joint is shown in Fig. 22, the dimensions of which are as follows:

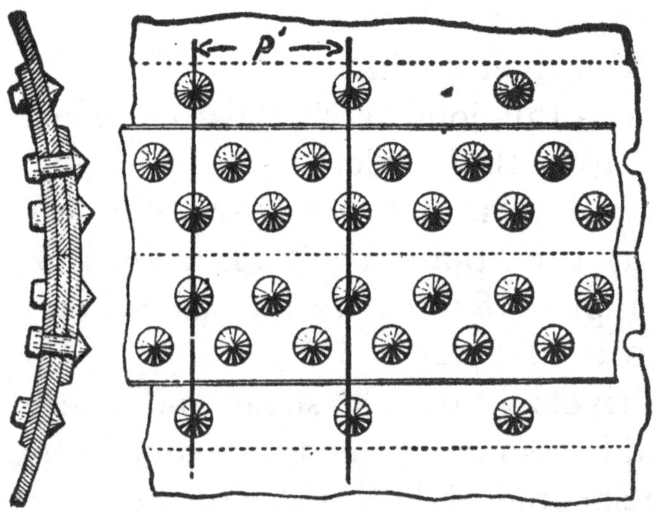

FIGURE 22

T = $\frac{7}{16}$ in.
D = $\frac{15}{16}$ in.
A = .69 in.
P = 3⅜ in.
P′ = 6¾ in.

Failure may occur in this joint in either one of five ways.

1. By tearing the plate at the outer row of rivets, where the pitch is 6¾ in. The net strength of the unit strip at this point is P′ − D × T × T.S., found as follows: 6.75 − .9375 × .4375 × 60,000 = 152,578.

2. By shearing four rivets in double shear and one in single shear. In this instance, of the four rivets in double shear, each one presents two sections, and the one in single shear presents one, thus making a total of nine sections of rivets to be sheared, and the strength is 9A × S, or .69 × 9 × 45,000 = 279,450.

3. Rupture of the plate at the middle row of rivets and shearing one rivet. Opposed to this strain the strength is P′ − 2D × T × T.S. + 1A × S, equivalent to 6.75 − (.9375 × 2) × .4375 × 60,000 + .69 × 90,000 = 190.068.

4. Crushing in front of four rivets and shearing one rivet. The resistance in this instance is 4D × T × C +

1A × S, or .9375 × 4 × .4375 × 90,000 + .69 × 45,000 = 178,706.

5. Failure may be caused by crushing in front of five rivets, four of which pass through both the inside and outside butt straps, while the fifth rivet passes through the inside strap only, and the resistance is 5D × T × C, equivalent to .9375 × 5 × 90,000 = 184,570.

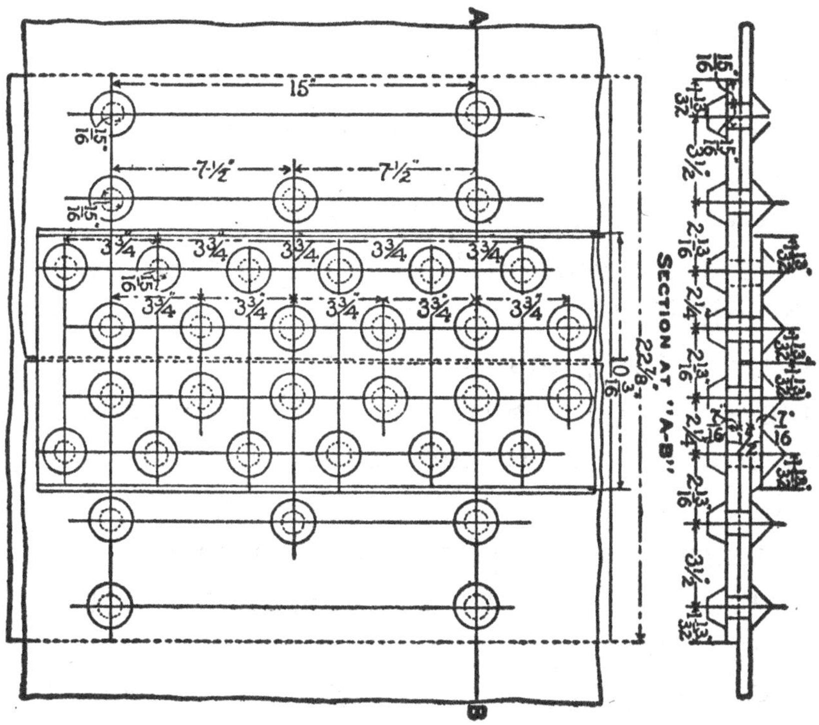

FIGURE 23

The strength of the unit strip of plate before drilling is P′ × T × T.S., or 6.75 × .4375 × 60,000 = 177,187, and the efficiency is 152,578 × 100 ÷ 177,187 = 86 per cent.

With the constantly increasing demand for higher steam pressures, the necessity for higher efficiencies in the riveted joints of boilers becomes more apparent, and of late years quadruple and even quintuple-riveted butt joints have in many instances come into use. The quadruple butt joint when properly designed shows a high efficiency, in some cases as high as 94.6 per cent. Fig. 23 illustrates a joint of this kind, and the dimensions are as follows:

$$T = \tfrac{1}{2} \text{ in.}$$
$$D = \tfrac{15}{16} \text{ in.}$$
$$A = .69 \text{ in.}$$
$$P, \text{ inner rows} = 3\tfrac{3}{4} \text{ in.}$$
$$P', \text{ 1st outer row} = 7\tfrac{1}{2} \text{ in.}$$
$$P'', \text{ 2d outer row} = 15 \text{ in.}$$

The two inner rows of rivets extend through the main plate and both the inside and outside cover plates or butt straps.

The two outer rows reach through the main plate and inside cover plate only, the first outer row having twice the pitch of the inner rows, and the second outer row has twice the pitch of the first.

Taking a strip or section of plate 15 in. wide (pitch of outer row), there are four ways in which this joint may fail.

1. By tearing of the plate at the outer row of rivets. The resistance is $P'' - D \times T \times T.S.$, or $15 - .9375 \times .5 \times 60{,}000 = 421{,}875$.

2. By shearing eight rivets in double shear and three in single shear. The strength in resistance is $19A \times S$, or $.69 \times 19 \times 45{,}000 = 589{,}950$.

3. By tearing at inner rows of rivets and shearing three rivets. The resistance is $P'' - 4D \times T \times T.S. + 3A \times S$, or $15 - (.9375 \times 4) \times .5 \times 60{,}000 + .69 \times 3 \times 45{,}000 = 430{,}650$.

4. By tearing at the first outer row of rivets, where the pitch is $7\tfrac{1}{2}$ in., and shearing one rivet. The resistance is $P'' - 2D \times T \times T.S. + A \times S$, or $15 - (.9375 \times 2) \times .5 \times 60{,}000 + .69 \times 45{,}000 = 424{,}800$.

It appears that the weakest part of the joint is at the

outer row of rivets, where the net strength is 421,875. The strength of the solid strip of plate 15 in. wide before drilling is P″ × T × T.S., or 15 × .5 × 60,000 = 450,000, and the efficiency is 421,875 × 100 ÷ 450,000 = 93.7 per cent.

Staying Flat Surfaces. The proper staying or bracing of all flat surfaces in steam boilers is a highly important problem, and while there are various methods of bracing resorted to, still, as Dr. Peabody says, "the staying of a flat surface consists essentially in holding it against pressure at a series of isolated points which are arranged in regular or symmetrical pattern." The cylindrical shell of a boiler does not need bracing, for the very simple reason that the internal pressure tends to keep it cylindrical. On the contrary, the internal pressure has a constant tendency to bulge out the flat surface. Rule 2, Section 6, of the rules of the U. S. Supervising Inspectors provides as follows: "No braces or stays hereafter to be employed in the construction of boilers shall be allowed a greater strain than 6,000 lbs per square inch of section."

The weakest portion of the crow foot brace when in position is at the foot end, where it is connected to the head by two rivets. With a correctly designed brace the pull on these rivets is direct and the tensile strength of the material needs to be considered only, but if the form of the brace is such as to bring the rivet holes above or below the center line of the brace, or if the rivets are pitched too far from the body of the brace, there will be a certain

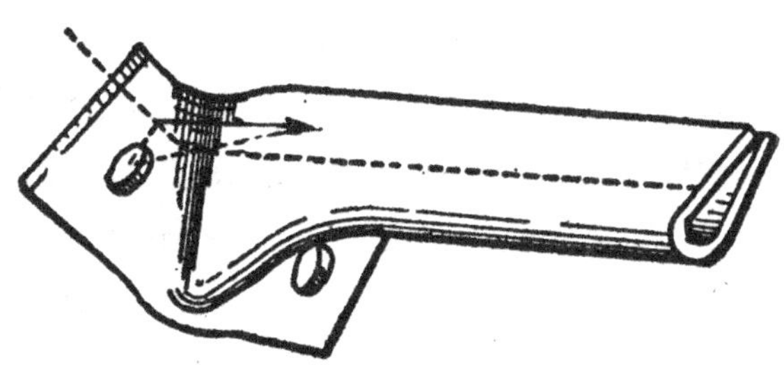

FIGURE 24

leverage exerted upon the rivets in addition to the direct pull. Fig. 24 shows a brace of incorrect design and Figs. 25 and 26 show braces designed along correct lines.

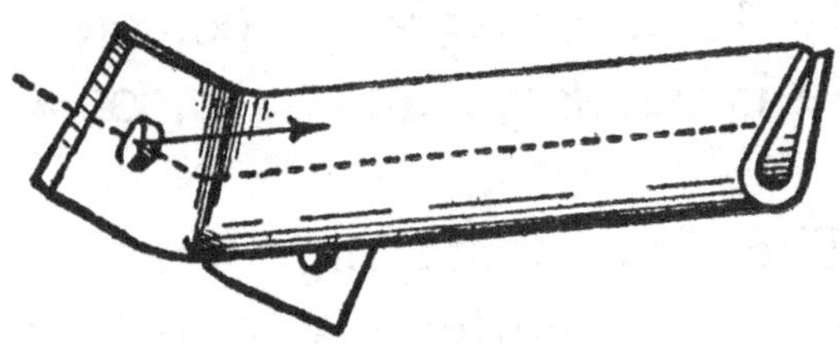

FIGURE 25

The problem of properly staying the flat crown sheet of a horizontal fire-box boiler, especially a locomotive boiler, is a very difficult one and has taxed the inventive genius of some of the most eminent engineers.

For simplicity of construction and great strength the cylindrical form of fire-box known as the Morison corrugated furnace has proved to be very successful. This form of fire-box was in 1899 applied to a locomotive by Mr. Cornelius Vanderbilt, at the time assistant superintendent of motive power of the New York Central and Hudson River R. R. This furnace was rolled of ¾-in. steel, is 59 in. internal diameter and 11 ft. 2¼ in. in length. It was tested under an external pressure of 500 lbs. per square inch before being placed in the boiler. It is carried at the front end by a row of radial sling stays from the outside plate, and supported at the rear by the back head. Figs. 27 and 28 show respectively a sectional view and an end elevation of this boiler. It will be seen at once that the question of stays for a fire-box of this type becomes very simple.

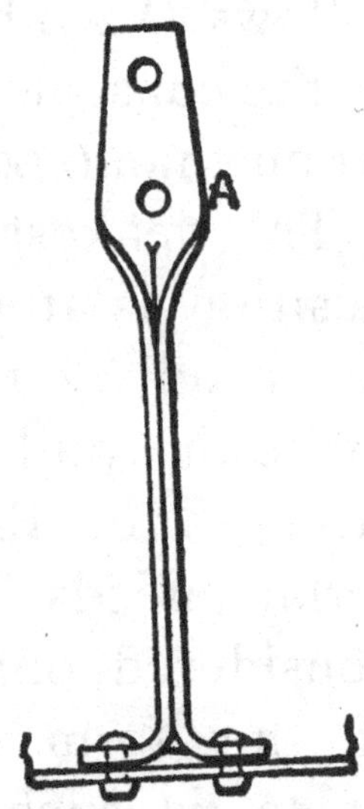

FIGURE 26

Calculating the Strength of Stayed Surfaces. In calculations for ascertaining the strength of stayed surfaces, or for finding the number of stays required for any

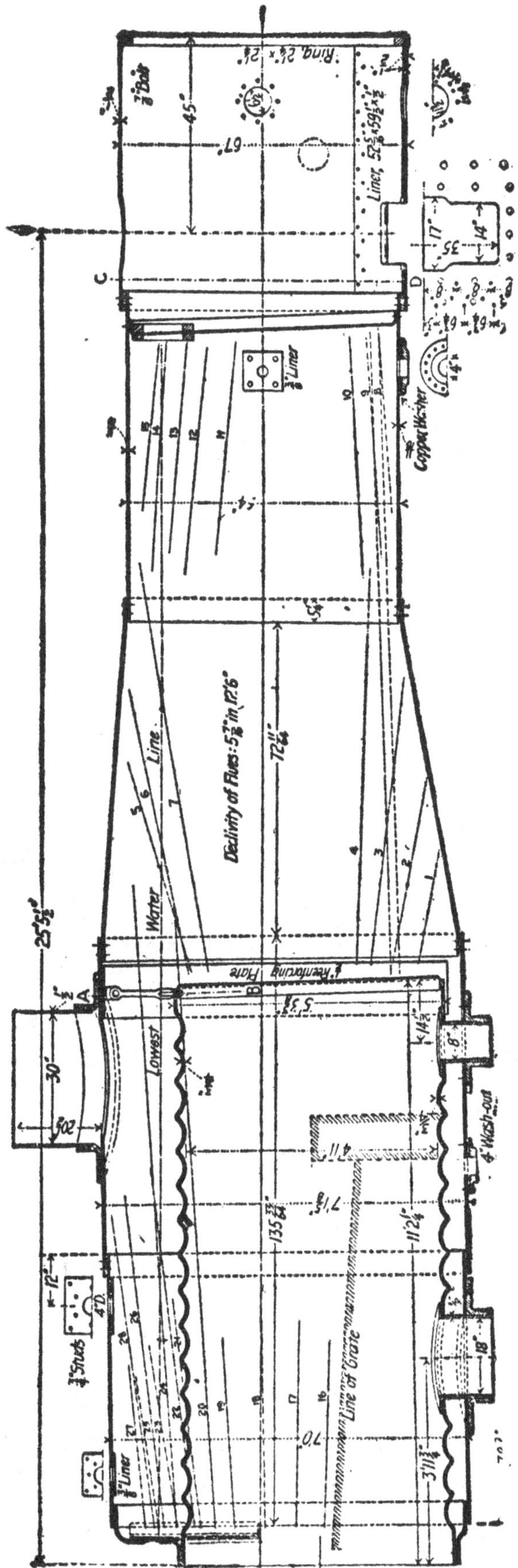
Declivity of Flues: 5 7/16" in 12'6"
Water Line
Lowest
Reinforcing Plate
Line of Grate
Wash-out
Copper Washer
Liner
Studs

FIGURE 27

given flat surface in a boiler, the working pressure being known, it must be remembered that each stay is subjected to the pressure on an area bounded by lines drawn midway between it and its neighbors. Therefore the area in square inches, of the surface to be supported by each stay, equals the square of the pitch

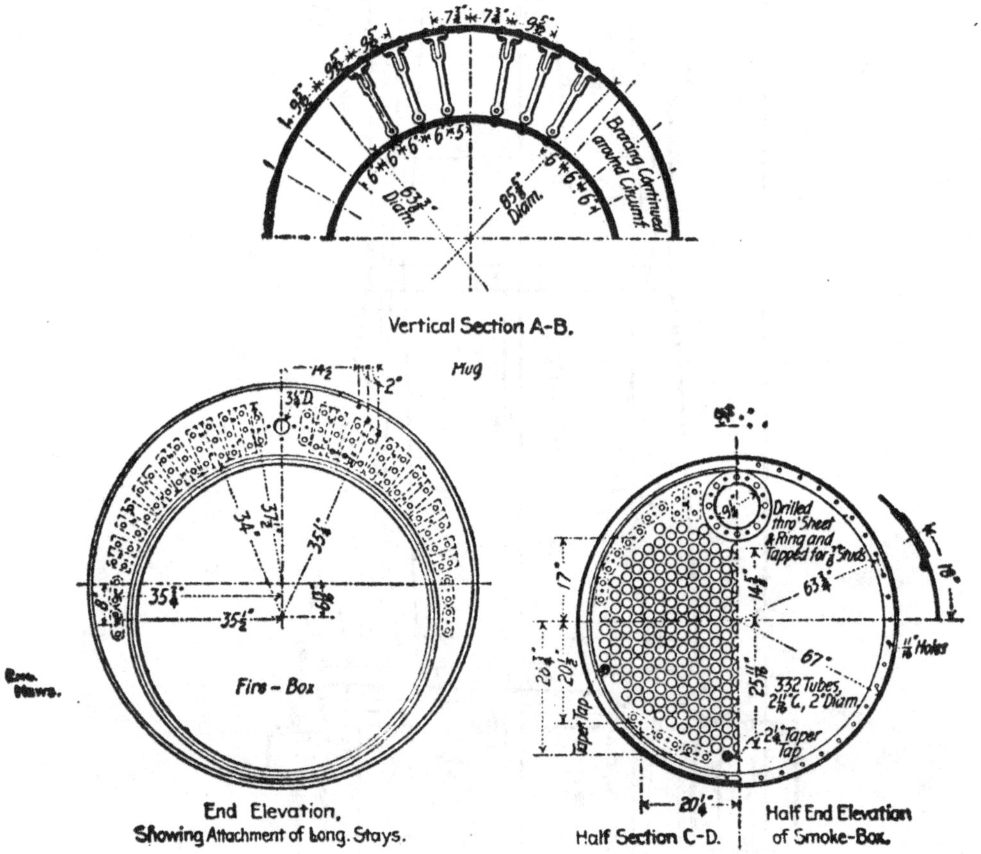

FIGURE 28

or distance in inches between centers of the points of connection of the stays to the flat plate. Thus, suppose the stays in a certain boiler are spaced 8 in. apart, the area sustained by each stay = 8 × 8 = 64 sq. in., or assume the stay bolts in a locomotive fire-box to be pitched 4½ in. each way, the area supported by each stay bolt = 4½ × 4½ = 20¼ sq. in.

The minimum factor of safety for stays, stay bolts

and braces is 8, and this factor should enter into all computations of the strength of stayed surfaces.

The pitch for stays depends upon the thickness of the plate to be supported, and the maximum pressure to be carried.

In computing the total area of the stayed surface it is safe to assume that the flange of the plate, where it is riveted to the shell, sufficiently strengthens the plate for a distance of 2 in. from the shell, also that the tubes act as stays for a space of 2 in. above the top row. Therefore the area of that portion of the flat head or plate bounded by an imaginary line drawn at a distance of 2 in. from the shell and the same distance from the last row of tubes is the area to be stayed. This surface may be in the form of a segment of a circle, as with a cylindrical boiler, or it may be rectangular in shape, as in the case of a locomotive or other fire-box boiler. Other forms of stayed surfaces are often encountered, but in general the rules applicable to segments or rectangular figures will suffice for ascertaining the areas.

By the use of Table 13 and the rule that follows, the area of the segmental portion of any boiler head may be ascertained.

Rule. Divide the height of the segment by the diameter of the circle. Then find the decimal opposite this ratio in the column headed "Area." Multiply this area by the square of the diameter. The result is the required area.

Example. Diameter of circle = 72 in. Height of segment = 25 in. 25 ÷ 72 = .347, which will be found in the column headed "Ratio," and the area opposite this .24212. Then .24212 × 72 × 72 = 1,255 sq. in., area of segment.

Table 13

Areas of Segments of a Circle

Ratio	Area	Ratio	Area	Ratio	Area	Ratio	Area
.2	.11182	.243	.14751	.286	.18542	.329	.22509
.201	.11262	.244	.14837	.287	.18633	.33	.22603
.202	.11343	.245	.14923	.288	.18723	.331	.22697
.203	.11423	.246	.15009	.289	.18814	.332	.22792
.204	.11504	.247	.15095	.29	.18905	.333	.22886
.205	.11584	.248	.15182	.291	.18996	.334	.22980
.206	.11665	.249	.15268	.292	.19086	.335	.23074
.207	.11746	.25	.15355	.293	.19177	.336	.23169
.208	.11827	.251	.15441	.294	.19268	.337	.23263
.209	.11908	.252	.15528	.295	.19360	.338	.23358
.21	.11990	.253	.15615	.296	.19451	.339	.23453
.211	.12071	.254	.15702	.297	.19542	.34	.23547
.212	.12153	.255	.15789	.298	.19634	.341	.23642
.213	.12235	.256	.15876	.299	.19725	.342	.23737
.214	.12317	.257	.15964	.3	.19817	.343	.23832
.215	.12399	.258	.16051	.301	.19908	.344	.23927
.216	.12481	.259	.16139	.302	.20000	.345	.24022
.217	.12563	.26	.16226	.303	.20092	.346	.24117
.218	.12646	.261	.16314	.304	.20184	.347	.24212
.219	.12729	.262	.16402	.305	.20276	.348	.24307
.22	.12811	.263	.16490	.306	.20368	.349	.24403
.221	.12894	.264	.16578	.307	.20460	.35	.24498
.222	.12977	.265	.16666	.308	.20553	.351	.24593
.223	.13060	.266	.16755	.309	.20645	.352	.24689
.224	.13144	.267	.16843	.31	.20738	.353	.24784
.225	.13227	.268	.16932	.311	.20830	.354	.24880
.226	.13311	.269	.17020	.312	.20923	.355	.24976
.227	.13395	.27	.17109	.313	.21015	.356	.25071
.228	.13478	.271	.17198	.314	.21108	.357	.25167
.229	.13562	.272	.17287	.315	.21201	.358	.25263
.23	.13646	.273	.17376	.316	.21294	.359	.25359
.231	.13731	.274	.17465	.317	.21387	.36	.25455
.232	.13815	.275	.17554	.318	.21480	.361	.25551
.233	.13900	.276	.17644	.319	.21573	.362	.25647
.234	.13984	.277	.17733	.32	.21667	.363	.25743
.235	.14069	.278	.17823	.321	.21760	.364	.25839
.236	.14154	.279	.17912	.322	.21853	.365	.25936
.237	.14239	.280	.18002	.323	.21947	.366	.26032
.238	.14324	.281	.18092	.324	.22040	.367	.26128
.239	.14409	.282	.18182	.325	.22134	.368	.26225
.24	.14494	.283	.18272	.326	.22228	.369	.26321
.241	.14580	.284	.18362	.327	.22322	.37	.26418
.242	.14666	.285	.18452	.328	.22415	.371	.26514

TABLE 13—*Continued*

Ratio	Area	Ratio	Area	Ratio	Area	Ratio	Area
.372	.26611	.405	.29827	.438	.33086	.471	.36373
.373	.26708	.406	.29926	.439	.33185	.472	.36471
.374	.26805	.407	.30024	.44	.33284	.473	.36571
.375	.26901	.408	.30122	.441	.33384	.474	.36671
.376	.26998	.409	.30220	.442	.33483	.475	.36771
.377	.27095	.41	.30319	.443	.33582	.476	.26871
.378	.27192	.411	.30417	.444	.33682	.477	.36971
.379	.27289	.412	.30516	.445	.33781	.478	.37071
.38	.27386	.413	.30614	.446	.33880	.479	.37171
.381	.27483	.414	.30712	.447	.33980	.48	.37270
.382	.27580	.415	.30811	.448	.34079	.481	.37370
.383	.27678	.416	.30910	.449	.34179	.482	.37470
.384	.27775	.417	.31008	.45	.34278	.483	.37570
.385	.27872	.418	.31107	.451	.34378	.484	.37670
.386	.27969	.419	.31205	.452	.34477	.485	.37770
.387	.28067	.42	.31304	.453	.34577	.486	.37870
.388	.28164	.421	.31403	.454	.34676	.487	.37970
.389	.28262	.422	.31502	.455	.34776	.488	.38070
.39	.28359	.423	.31600	.456	.34876	.489	.38170
.391	.28457	.424	.31699	.457	.34975	.49	.38270
.392	.28554	.425	.31798	.458	.35075	.491	.38370
.393	.28652	.426	.31897	.459	.35175	.492	.38470
.394	.28750	.427	.31996	.46	.35274	.493	.38570
.395	.28848	.428	.32095	.461	.35374	.494	.38670
.396	.28945	.429	.32194	.462	.35474	.495	.38770
.397	.29043	.43	.32293	.463	.35573	.496	.38870
.398	.29141	.431	.32392	.464	.35673	.497	.38970
.399	.29239	.432	.32941	.465	.35773	.498	.39070
.4	.29337	.433	.32590	.466	.35873	.499	.39170
.401	.29435	.434	.32689	.467	.35972	.5	.39270
.402	.29533	.435	.32788	.468	.36072		
.403	.29631	.436	.32887	.469	.36172		
.404	.29729	.437	.32987	.47	.36272		

Strength of Unstayed Surfaces. A simple rule for finding the bursting pressure of unstayed flat surfaces is that of Mr. Nichols, published in the Locomotive, February, 1890, and quoted by Professor Kent in his pocket-book. The rule is as follows: "Multiply the thickness of the plate in inches by ten times the tensile strength of the material used, and divide the product by the area of the head in square inches." Thus:

Diameter of head = 66 in.
Thickness of head = 5/8 in.
Tensile strength = 55,000 lbs.
Area of head = 3,421 sq. in.

$5/8 \times 55{,}000 \times 10 \div 3{,}421 = 100$, which is the number of pounds pressure per square inch under which the unstayed head would bulge.

If we use a factor of safety of 8, the safe working pressure would be $100 \div 8 = 12.5$ lbs. per square inch, but as the strength of the unstayed head is at best an uncertain quantity it has not been considered in the foregoing calculations for bracing, except as regards that portion of it that is strengthened by the flange.

In all calculations for the strength of stayed surfaces, and especially where diagonal crow foot stays are used, the strength of the rivets connecting the stay to the flat plate must be carefully considered. A large factor of safety, never less than 8, should be used, and the cross section of that portion of the foot of the stay through which the rivet holes are drilled should be large enough, after deducting the diameter of the hole, to equal the sectional area of the body of the stay.

Dished Heads. In boiler work where it is possible to use dished, or "bumped up" heads as they are sometimes called, this type of head is rapidly coming into use. Dished heads may be used in the construction of steam drums, also in many cases for dome-covers, thus obviating the necessity of bracing.

As there has been a constantly growing demand for an increase in the power of locomotives, and as the boiler is the source of power, builders have been constrained to change the design of locomotive boilers in such manner as would bring about an enlargement of both the heating surface and the grate area. Consequently the old wagon top type of boiler, with the fire-box down between the drivers and close to the track, has been largely superseded by the modern straight-top boiler having a wide fire-box, which as applied to freight engines with low wheels is usually above the rear drivers, but as applied to passenger engines with high wheels is usually behind the rear drivers and supported by trailing wheels, as in "Atlantic 4-4-2," "Prairie 2-6-2" and the "Pacific 4-6-2" types. The introduction of the wide fire-box and consequent increase of great area has made it possible to

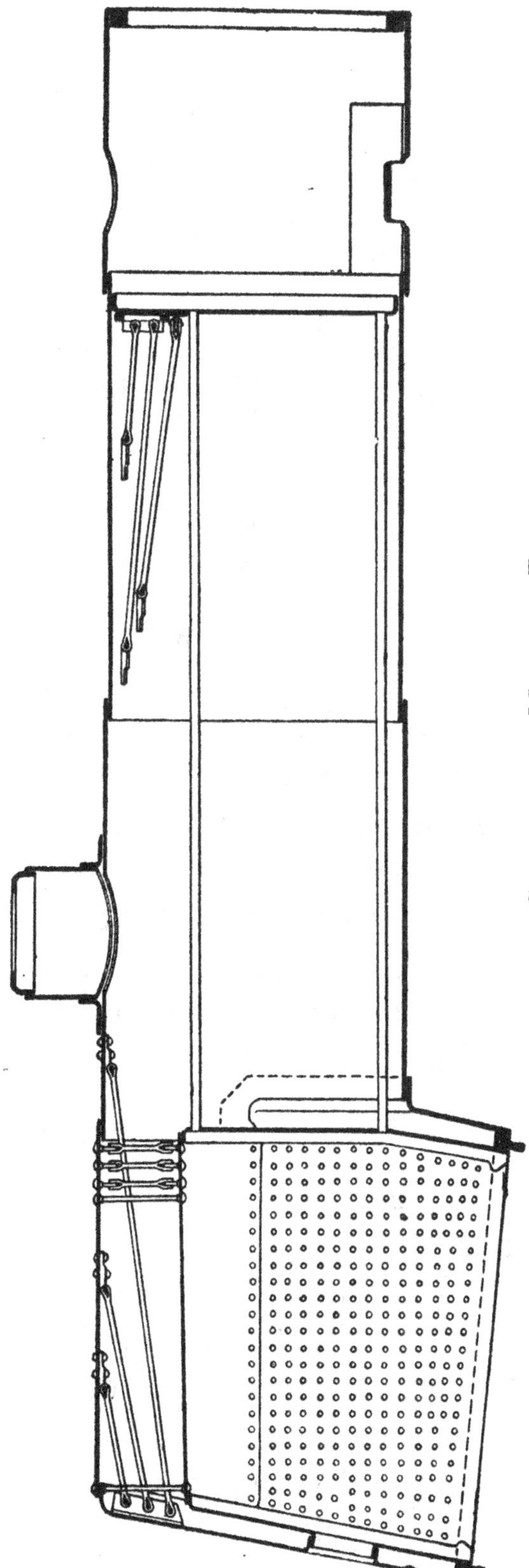

FIGURE 29. SECTION OF MODERN BOILER

burn cheaper grades of coal than was possible with the older type of boiler. It may be used (with some modifications) for both soft and hard coal.

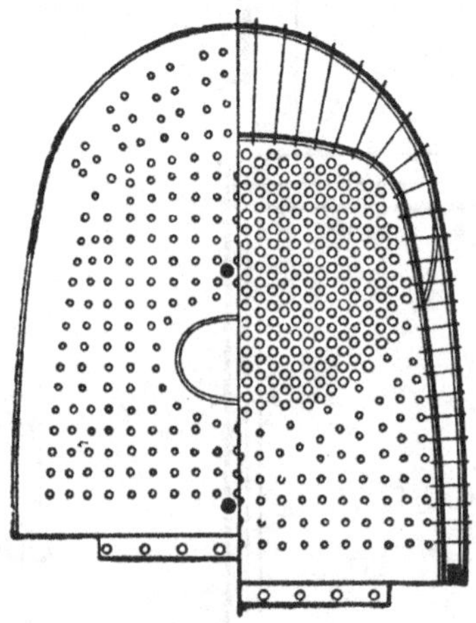

FIGURE 30

Fig. 29 shows a sectional elevation of a modern locomotive boiler, and Fig. 30 an end view of one-half of the flue sheet and one-half of the back head.

The staying of the heads and crown sheet is clearly illustrated. The general dimensions of the fire-box at the present time varies from 8 ft. to 10 ft. 4 in. in length, with a width of from 40 to 42 in., and a depth of 6 to 7 ft. in front, and 5 ft. 6 in. to 6 ft. 6 in. at the back, the size depending upon the type of engine and the kind of work it was designed to perform.

The diameter of the barrel or cylindrical portion of locomotive boilers built for train service varies all the way from 60 to 78 in., and some recent splendid examples of the locomotive builders' art have boilers 83 in. in diameter.

QUESTIONS

64. What are the four vital organs of a locomotive boiler?

65. Describe the mud ring.

66. Describe in general terms the fire-box.

67. How are the sides of the fire-box stayed?

68. Describe a stay bolt.

69. How far apart, center to center, are stay bolts usually spaced?

70. What causes stay bolts to break?

71. Why are stay bolts made hollow?

72. Describe the flexible stay bolt.

73. What advantage has a flexible stay bolt over a rigid one?

74. Is it necessary to strengthen the crown sheet by stays?

75. Why does the crown sheet need to be supported?

76. Name the three methods usually employed for staying the crown sheet.

77. Describe crown bars, and how applied.

78. Why is there a space preserved between the crown bars and top of crown sheet?

79. How are the crown bolts attached?

80. Why are thimbles placed between the crovn bars and top of crown sheet?

81. Describe the radial system of staying the crown sheet.

82. What is the principal defect in this system?

83. Describe the Belpaire system.

84. What great advantage has this form of fire-box over others?

85. Why are crown sheets usually made to slope downwards from the front to the back end?

86. How are the heads of the boiler usually stayed?

87. What are diagonal crow foot stays?

88. How is the flue sheet braced?

89. What is a gusset stay, and how is it connected to the head and shell?

90. Why should the flue sheet be thicker than the shell?

91. What advantage is there in setting the tubes in vertical rows?

92. What is the usual diameter of locomotive tubes?

93. How are the tubes made water-tight in the sheet?

94. Describe the Prosser tube expander and method of using it.

95. Describe the Dudgeon roller expander.

96. How is it used?

97. What is meant by the expression tensile strength of a boiler sheet?

98. What is the usual tensile strength of steel boiler plate?

99. What should be the tensile strength of the rods from which rivets are made?

100. What is the shearing resistance of iron rivets?

101. What is the shearing resistance of steel rivets?

102. What is meant by the efficiency of a riveted joint?

103. What type of joint should be used for plates ½ in. thick or more?

104. Give the diameter of rivet pitch, and efficiency of a double-riveted joint.

105. What is the usual efficiency of single-riveted joints?

106. How should double-riveted joints be designed in order to obtain the highest efficiency?

107. Describe a triple-riveted butt joint.

108. How should a triple-riveted butt joint be designed in order to obtain the highest efficiency?

109. What is meant by the expression, the unit strip or net section of plate, as used in calculating the efficiency of a riveted joint?

110. What is the usual efficiency of the triple-riveted butt joint?

111. What efficiency per cent does the quadruple-riveted butt joint show when properly designed?

112. Why is it that the cylindrical portion of a boiler does not require to be stayed?

113. What effect does the pressure inside a boiler have upon flat surfaces, such as the heads, crown sheet, etc.?

114. Where is the weakest portion of a crow foot brace?

115. How is the area in square inches to be supported by each stay ascertained?

116. What is the minimum factor for stays and stay bolts?

117. What two factors govern the pitch for stays?

118. What portions of the heads do not need to be braced?

119. Is it possible to weld boiler seams?

120. Describe in general terms the modern locomotive boiler.

121. What are the general dimensions of the firebox?

CHAPTER III

THROTTLE AND DRY PIPE

Having studied at some length the construction of the boiler and the generation of steam, it is now in order to examine into the method by which the steam is conveyed to the cylinders of the engine, where it, or rather the heat that it contains, performs its work. The main factors in the transmission of the steam from the boiler to the interior of the cylinders, and from there to the open air, are the throttle valve and pipe, the dry pipe, the steam pipes and passages, the valves and ports, the exhaust passages and ports, and the exhaust nozzles. These will each be described in regular order, with the exception of the valves and ports, which will be fully described in the chapters on valves and valve setting.

The steam dome O, Fig. 6, is a cylindrical chamber made of boiler plate and riveted to the top of the boiler, usually directly over the fire-box. The function of the dome is to serve as a steam chamber that is elevated as high as possible above the surface of the water in the boiler, in order that the steam supplied to the cylinders, all of which is drawn from this chamber, may be as dry as it is possible to have it.

The steam is conducted from the dome to the cylinders through the dry pipe P-Q-R, Fig. 6, which extends from the top of the dome to the front flue sheet or head of the boiler. Connected to the front end of the dry pipe, inside the smoke box, are two cast iron curved pipes 1-2, Fig. 31, called the steam pipes, which conduct the steam to the steam chests, or valve chests as they are sometimes called. The horizontal portion of the dry pipe extending through the boiler is

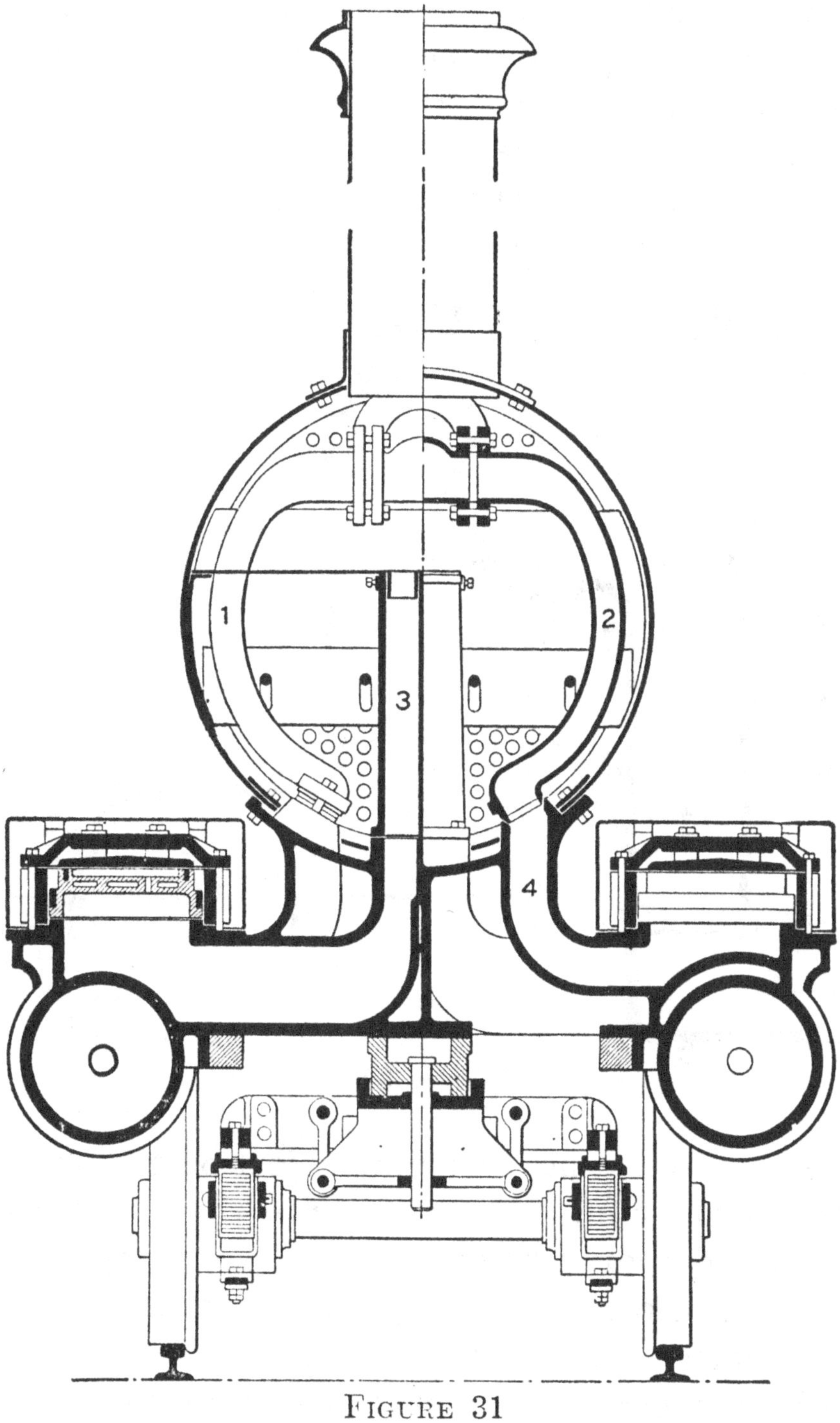

FIGURE 31

made of wrought iron, and the vertical portion T, Fig. 6, called the throttle pipe, and which is within the dome, is made of cast iron. At the top end of this pipe, near the top of the dome, the throttle U, Fig. 6, for controlling the steam, is usually located, although not always, as it is sometimes placed in the smoke box at the front end R of the dry pipe.

Formerly the throttle valve was a plain slide valve that moved upon a seat in which were ports similar in form to the steam ports in the valve chests, but

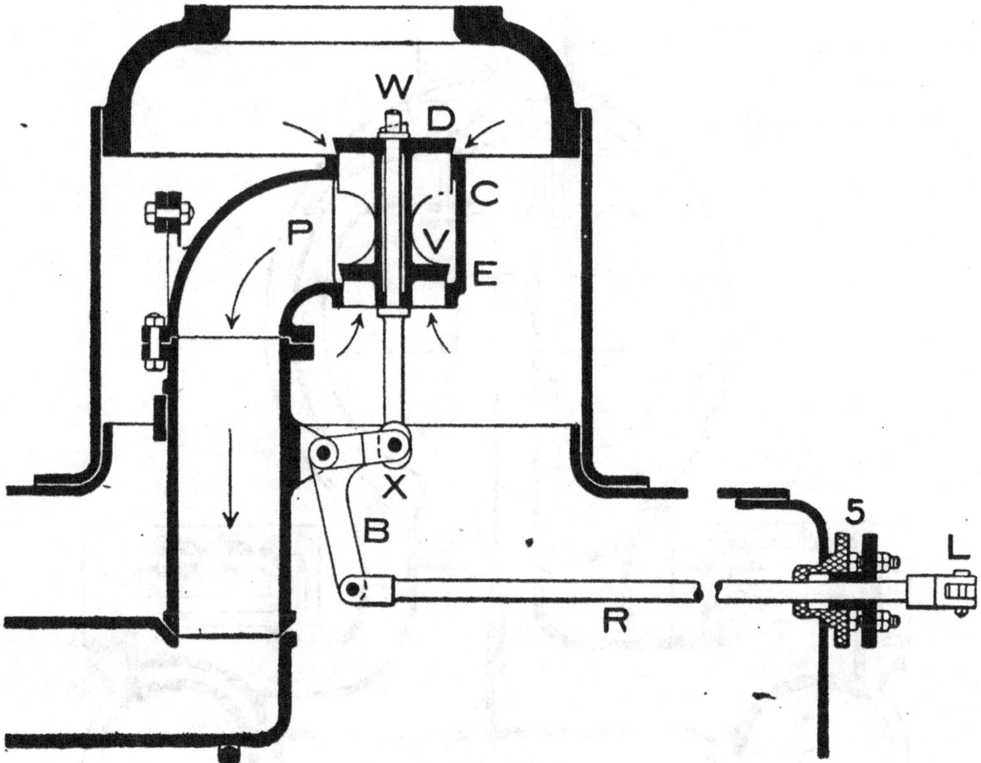

Figure 32

smaller in size. The principal objection to this type of throttle valve for a locomotive was that the pressure of the steam upon it when closed made it very difficult to open the throttle gradually, or to regulate or adjust it while open—two very important points in the operation of a locomotive. A much better form of throttle has been largely adopted in late years. This valve is shown at U, Fig. 6, and on a larger scale by Figs. 32

and 33, which give a sectional view and a plan of the throttle pipe, valve, and throttle lever.

The valve V, Fig. 32, is a double poppet valve, having two circular disks D and E, which cover two corresponding openings in the case C on the end of the pipe P. When the valve is raised and the disks are off their seats the steam flows in around their edges, as shown by the arrows. The disks are not the same

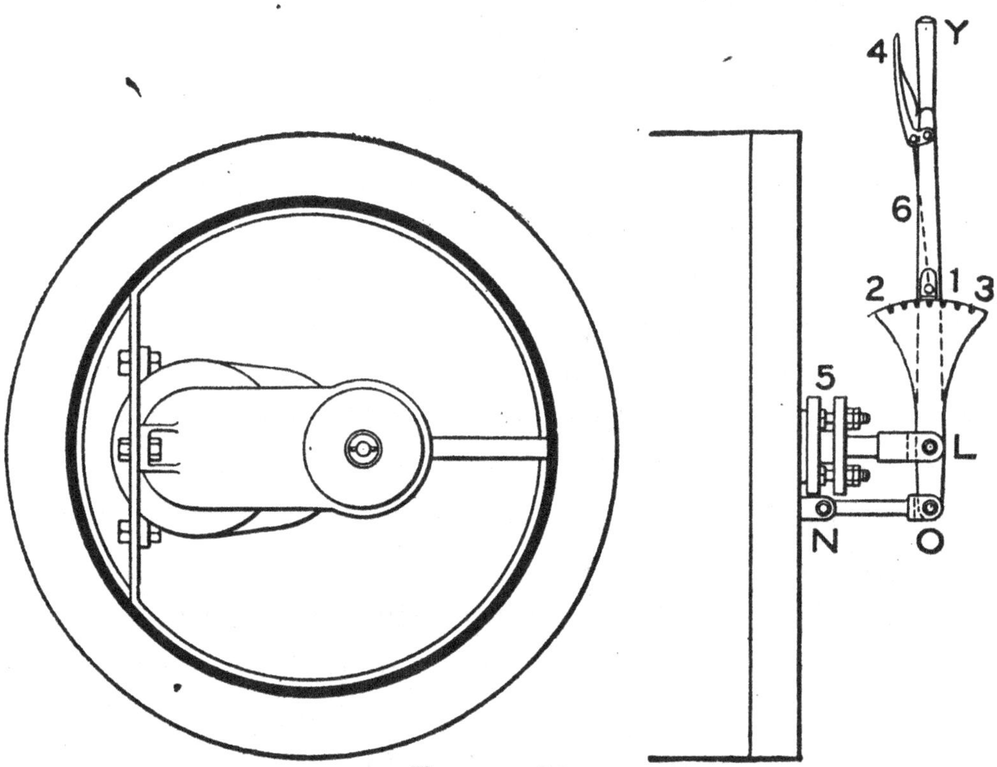

FIGURE 33

diameter, the top one being slightly larger. The steam pressure in the boiler acts upon the top of disk D and upon the bottom of disk E. If the two disks were exactly the same in diameter the valve would be balanced, but this is not desirable, as there might thus be a possibility of its being opened accidentally after the engineer had closed it. There is also another reason why the lower disk must be smaller in diameter than the upper one, viz., that it may be introduced through the top opening of the casing C, so as to cover the lower opening. There is thus a slightly greater pressure on the top surface of the upper disk tending to keep the valve closed, than there is on the bottom surface of the lower disk tending to raise the valve and open it. This arrangement of the parts causes the throttle to stay in any position it may be placed, while at the same time it moves comparatively easily. The means whereby the throttle is opened and closed are also shown in Figs. 32 and 33.

The stem W-X of the valve V extends downwards and connects with the upper arm of the bell crank B, Fig. 32. Connected to the lower arm of this bell crank, and extending through the back boiler head into the cab, is a rod R, called the throttle stem. This rod passes through a steam-tight stuffing box in the boiler head. The throttle lever Y, Fig. 33, is connected to the throttle stem at L and attached to a link N-O at O. This link is connected to the boiler head by a stud and pin at N, Fig. 33. The link is free to vibrate slightly, which enables the connection at L to move in a straight line. This provision causes the stem R, Fig. 32, to also move in a straight line in the stuffing box 5, which is very necessary in order that it may be kept steam-tight. Referring to Fig. 33, which is a plan view, the throttle lever Y is fitted with a latch 1 that gears into the curved rack 2-3, in order to hold the throttle in any required position. The latch 1 is operated by a trigger 4, connected by the rod.

The steam, being admitted by the throttle valve V into the throttle pipe P, passes on into the dry pipe P-Q-R, Fig. 6. This pipe, after passing through the front flue sheet of the boiler, is fitted with a T-pipe, thus dividing it into two branches to which the steam pipes are connected. These connections, which are all within the smoke-box, are clearly illustrated in Fig. 31, to which reference is now made.

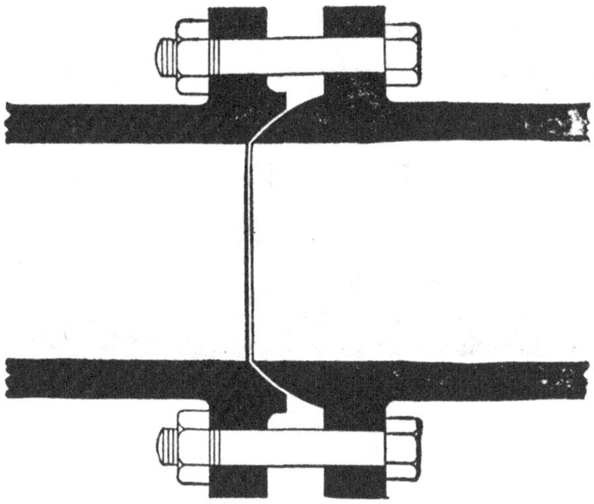

FIGURE 34

The steam pipes 1 and 2 are connected to each of the two branches of the T pipe at their top ends and to the cylinder castings at their bottom ends. The steam is thus conducted to the valve chests. Fig. 31 shows a sectional view of one of the steam pipes, 2 on the right and a section of one of the exhaust pipes 3 on the left. The steam pipes are exposed to great changes of temperature as a result of their being within the smoke-box, and consequently the wide range of expansion and contraction to which they are subjected renders it very difficult to keep the joints tight.

Another difficulty is also generally encountered in the assembling of the various parts forming these connections, as, for instance, if the upper end of pipe 4 in the cylinder casting, Fig. 31, were either too near or too far from the center line of the engine it would be necessary to move the end of pipe 2, either to the right or to the left, in order to bring it in line for connecting to 4. It is therefore necessary that there be a certain degree of flexibility in these connections, and this is accomplished by the use of ball joints. Fig. 34 illustrates a ball joint. The end of one of the pipes is turned into the form of a sphere or globe, and the end of the other pipe is formed into a corresponding concave shape, as shown in Fig. 34. This form of joint permits a lateral movement in either direction of the lower end of pipe 2 to bring it in line with the upper end of pipe 4.

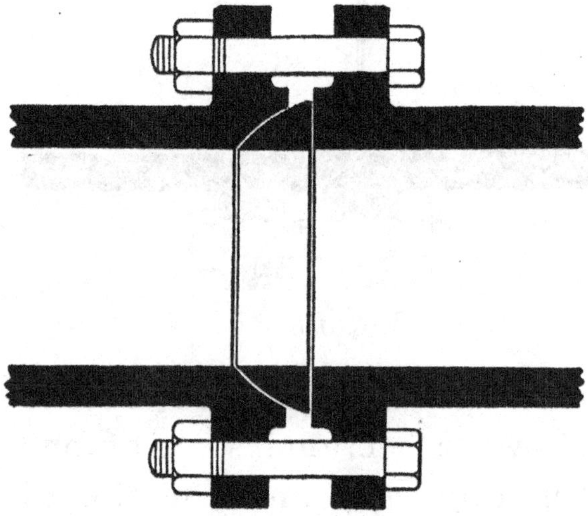

FIGURE 35

Another and still better form of flexible joint is illustrated in Fig. 35. In this joint a ring is interposed between the ends of the pipes. One side of this ring is spherical and the other side is flat, the ends of the pipes being shaped to correspond. With this form of joint the pipes are slightly adjustable in every direction, and the joints accommodate themselves to any and all motion that may be caused by expansion and contraction.

The exhaust pipes or nozzles are made of cast iron. Sometimes a single nozzle is used, such as shown in section in Fig. 36, having a partition at its base. In other cases two nozzles are used, which are generally cast together, as shown in section in Fig. 37.

Fig. 38 is a plan view of single and double nozzles. Rings or bushings are fitted in the outlet openings of these nozzles for the purpose of reducing their area and thereby increasing the draft. These bushings are made of various diameters and are easily removed in order to substitute others with larger or smaller openings as they may be required. If the exhaust orifice is too large the draught through the tubes will not be sufficient. On the other hand, if the area of the exhaust opening is reduced too much the back pressure in the cylinders will be increased, thereby limiting the power of the engine. It is therefore necessary that

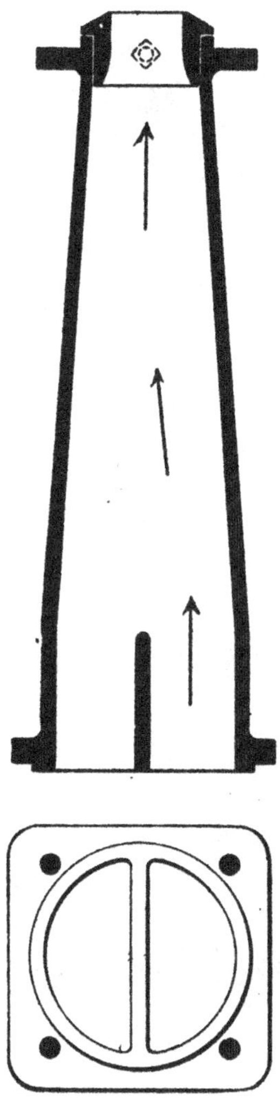

FIGURE 36

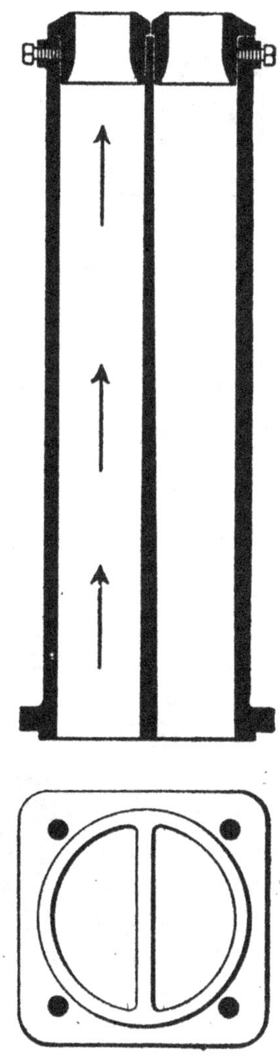

FIGURE 37

great care and good judgment be exercised in the adjustment of the exhaust nozzles.

Various devices have been invented for adjusting the area of the exhaust nozzles while the engine is working steam, but none has proved to be satisfactory, and the old method of adjustment when the engine is not working is still in vogue. A few of the many devices that have been invented for regulating the draft will be described in this connection.

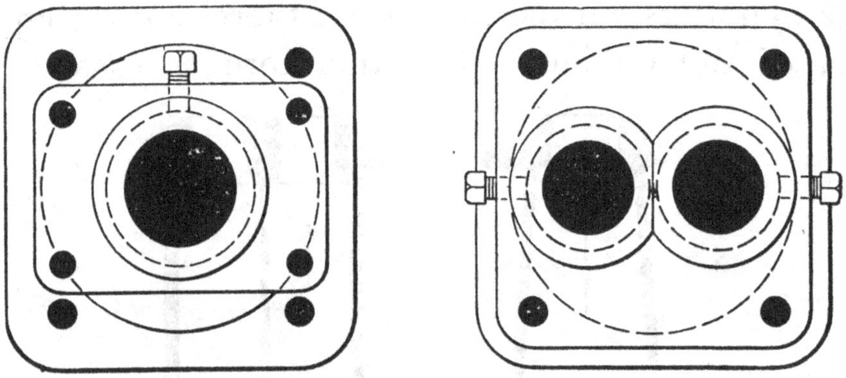

FIGURE 38

Fig. 39 shows a form of adjustable nozzle that appears to have considerable merit. It is the inven-

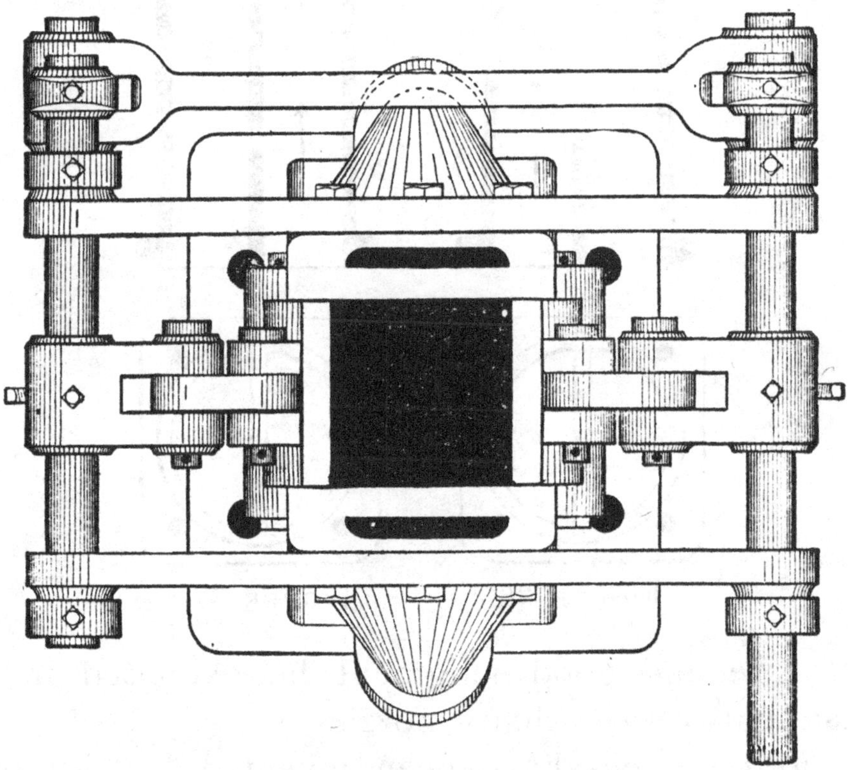

FIGURE 39

tion of Messrs. Wallace and Kellog, two engineers on the St. P., M. and O. R. R., and it has been used to some extent on that road, also on the Duluth and Iron Range R. R. The device is automatic in its operation, the regulating mechanism being connected to the reverse lever, or the reach rod, in such a manner that as the lever is moved from the center notch towards either corner the area of the nozzle is increased one-half square inch for each notch. It may be set so that with the reverse lever in either corner there will be seven square inches more of nozzle area than there is with the lever in or near the center notch.

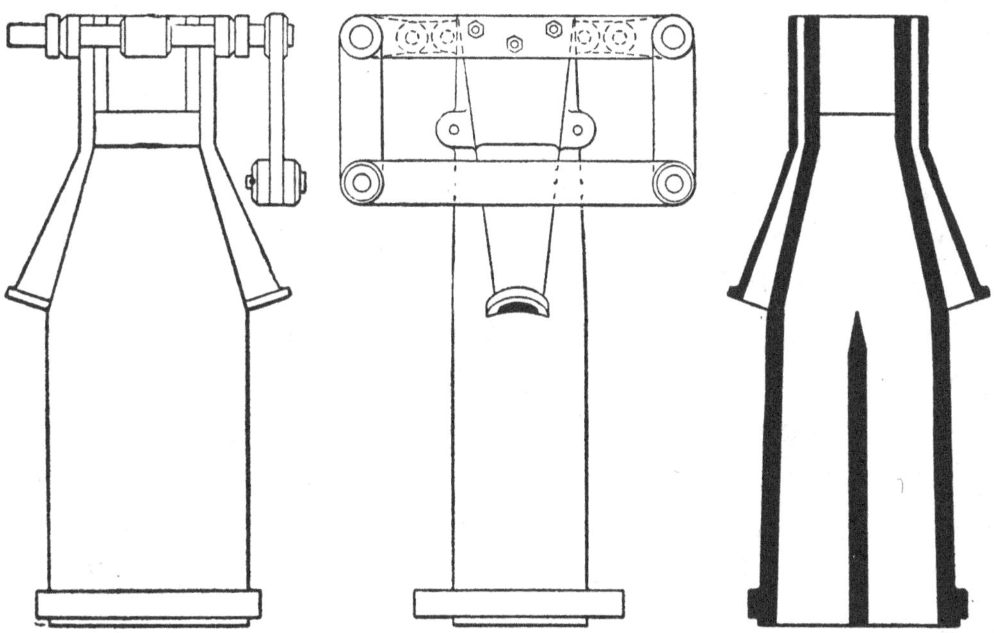

FIGURE 40

The nozzle areas for different positions of the reverse lever are as follows: Center notch, 22 sq. in.; second notch, 23 sq. in.; fourth notch, 24⅜ sq. in.; sixth notch, 25½ sq. in.; eight notch, 26$\frac{9}{16}$ sq. in.; tenth notch, 28½ sq. in., and in the corner, 29$\frac{11}{16}$ sq. in. The device is said to work satisfactorily and has shown a saving in fuel of from $59.00 to $97.00 per month over the ordinary nozzle.

Fig. 39 shows a plan and Fig. 40 an elevation, the cuts being self-explanatory.

The nozzle itself is square, and the adjustment is caused by two hinged ears which open as the reverse lever is moved from center towards corner and close as the lever is hooked back towards the center notch, so that the more steam that is being used the larger will be the nozzle area, and vice versa.

The De Lancey Exhaust Nozzle. This is another form of variable exhaust nozzle, as may be seen by the illustrations. It is the invention of Mr. John J. De Lancey, of Binghamton, N. Y., who describes his device in the following words:

"The object of my invention is to provide a new and improved exhaust nozzle for locomotives, serving to regulate the exhaust of the engines, and thereby regulating the draft in the boiler.

"Fig. 41 is a side elevation of the improvement as applied to a locomotive, parts being broken out. Fig. 42 is an enlarged plan view of the improvement. Fig. 43 is a transverse vertical section of the same. Fig. 44 is a sectional side elevation of the same on the line *x x* of Fig. 42, and Fig. 45 is a plan view of a modified form of the plate.

"The improved exhaust nozzle is provided with a plate B, fitted onto the upper end of the exhaust pipe C, which may be double, as is illustrated in Fig. 43, or single—that is, the two exhausts of the engines of the locomotive running into a single exhaust pipe.

"The plate B is provided with apertures D of the same size as the apertures at the upper ends of the exhaust pipe C, so that when the plate B is in a central or normal position the apertures D of the plate B fully register with the openings in the end of the exhaust

THROTTLE AND DRY PIPE

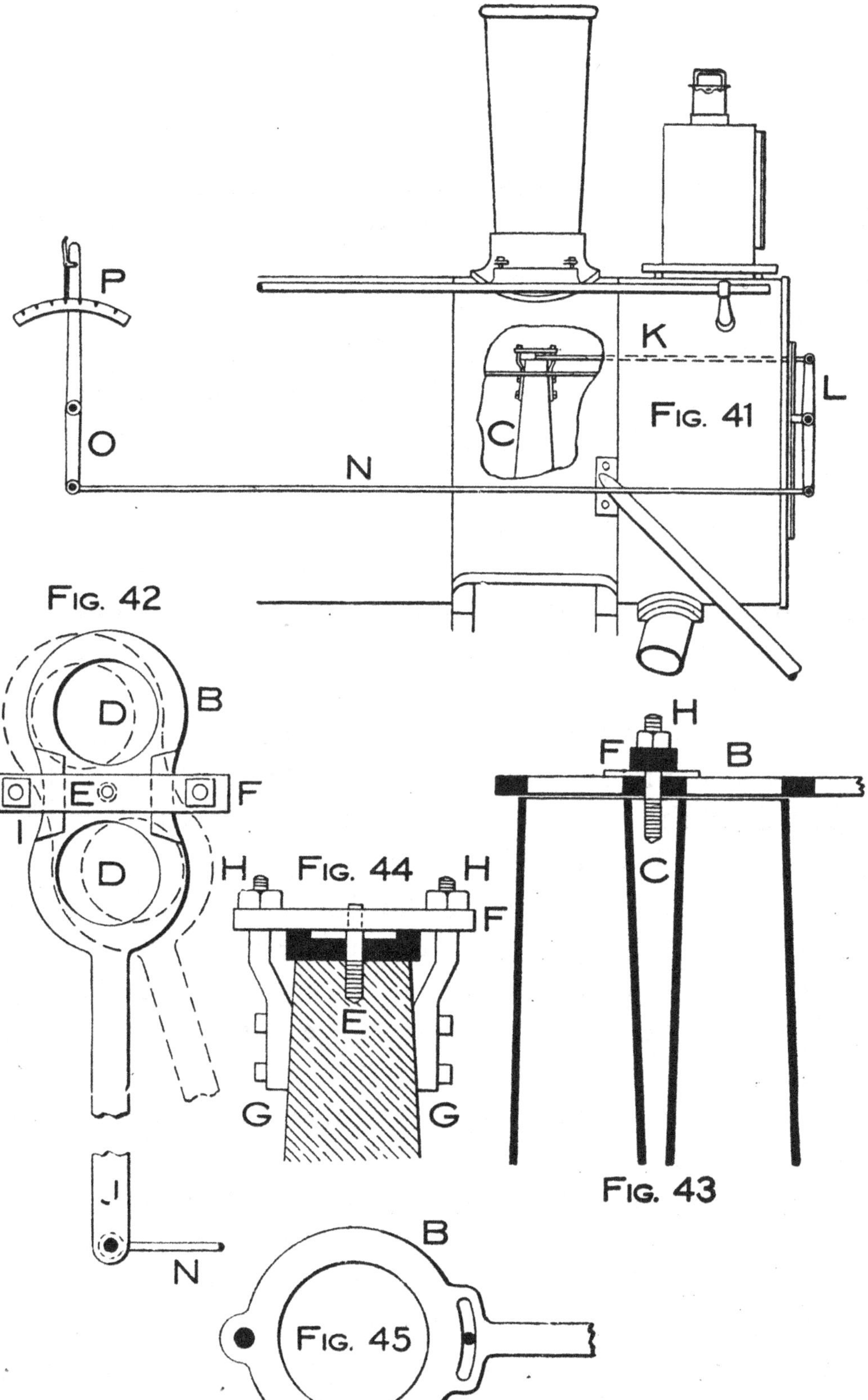

Figures 41 to 45

nozzle. The plate B is fulcrumed in its middle on a pin E, projecting from a bar F, supported on brackets G, secured to the sides of the exhaust pipe C, the said plate being held in place on the brackets by nuts H, screwing on the threaded ends of the said brackets G, as is plainly illustrated in Fig. 44. The pin E, after passing through the plate B, also passes a short distance into the top of the exhaust pipe C, so as to form a secure bearing for the plate B. On the top of the latter, at its sides in the middle, are arranged offsets I, onto which fits the under side of part of the bar F in such a manner that the plate B is free to turn on its pivot E, and at the same time is held securely against the upper end of the exhaust pipe C to prevent the plate from being lifted upward by the force of the exhaust steam.

"From the plate B projects to one side an arm J, pivotally connected by a link K with a lever L, fulcrumed on the outside at the front end of the locomotive boiler, the link K passing through the said front end. The lever L is also pivotally connected by a link N, extending along the outside of the locomotive, with a lever O, pivoted on the cab of the locomotive and extending into the same so as to be within convenient reach of the engineer in charge of the locomotive. The lever O is adapted to be locked in place in any desired position by the usual arrangement connected with a notched segment P, as shown in Fig. 41.

"When the lever O stands in a vertical position, as illustrated in the said figure, the openings D in the plate B fully register with the openings in the exhaust pipe C. In this position the exhaust steam can pass freely out of the exhaust pipe C through the smokebox and smokestack of the locomotive, so as to cause considerable draft in the fire-box of the boiler. When it is desirable to increase the amount of draft in the fire-box of the locomotive, the engineer in charge of the locomotive operates the lever O either forward or backward, so that the lever L swings and imparts a swinging motion by the link K and the arm J to the plate B, which latter moves across the top of the exhaust pipe C, and part of the openings of the latter are cut off or diminished in size, so that the exhaust of the engine is retarded, and consequently the draft in the smoke-box and smokestack is increased, so that a consequent increase of the draft takes place in the fire-box of the locomotive.

"It will be seen that the two openings in the exhaust pipe are diminished in size alike by moving the plate B, and it is immaterial in which direction the engineer moves the lever O, as the cut-off takes place either way."

Fig. 46 shows the Canby draft regulating apparatus, invented by Mr. Joseph C. Canby of Orange, Luzerne Co., Pa., and the following description of the device is furnished by the inventor himself:

"My invention relates to draft-regulating apparatus for locomotive and that class of boilers; and it consists of a smokestack with an adjustable petticoat or mouthpiece to equalize the draft through all the flues, also an arrangement of pipes and valves to introduce fresh air into the smokestack to check the draft without opening the fire door and letting the cold air in onto the boiler and tubes, thereby making a great saving in the fuel and being better for the boiler and flues."

Fig. 46 represents the front view of the boiler with the automatic draft-regulator attached. Fig. 47 is a horizontal section of front of boiler, showing smokestack and rock shaft. Fig. 48 is a longitudinal section of the smoke-box and boiler, showing the connection of the valve N and regulator O and the connection of arm J to the cab K by the rod R.

A B C represent the sections of the smokestack, or, as familiarly called, "petticoats," arranged with lugs

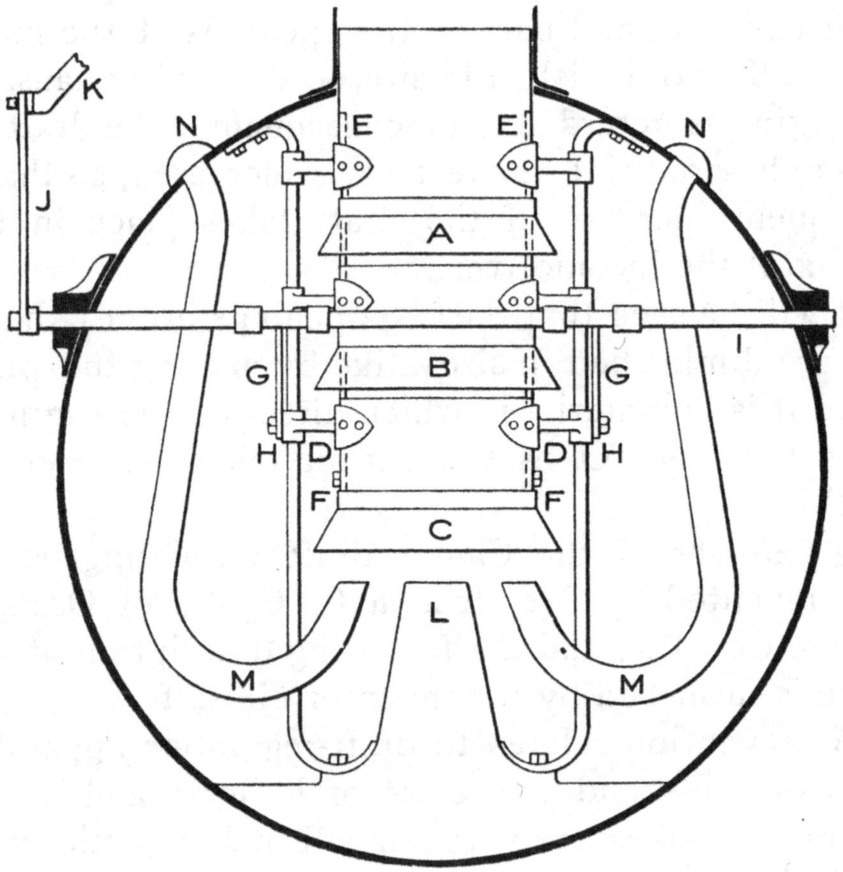

FIGURE 46

D on the sides with slides E E, having slots and set screws F F, by which they are adjusted to the space required between them, thereby enabling the engineer to equalize the draft in the fire-box, as experience shows that when the draft is nearest to the bottom of the smoke jacket the draft is strongest on the back end of the fire next the flue, and by decreasing there and increasing it in the top flues the draft is made stronger in the front part of the fire-box. This more nearly equalizes the combustion of the fuel. The connecting rods G G are attached to the

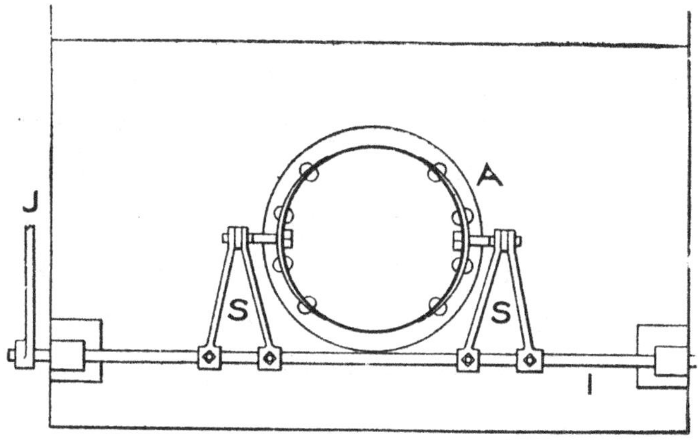

FIGURE 47

lugs H, and the arms S S project from the rocking shaft I, which is operated by the arm J and rod K,

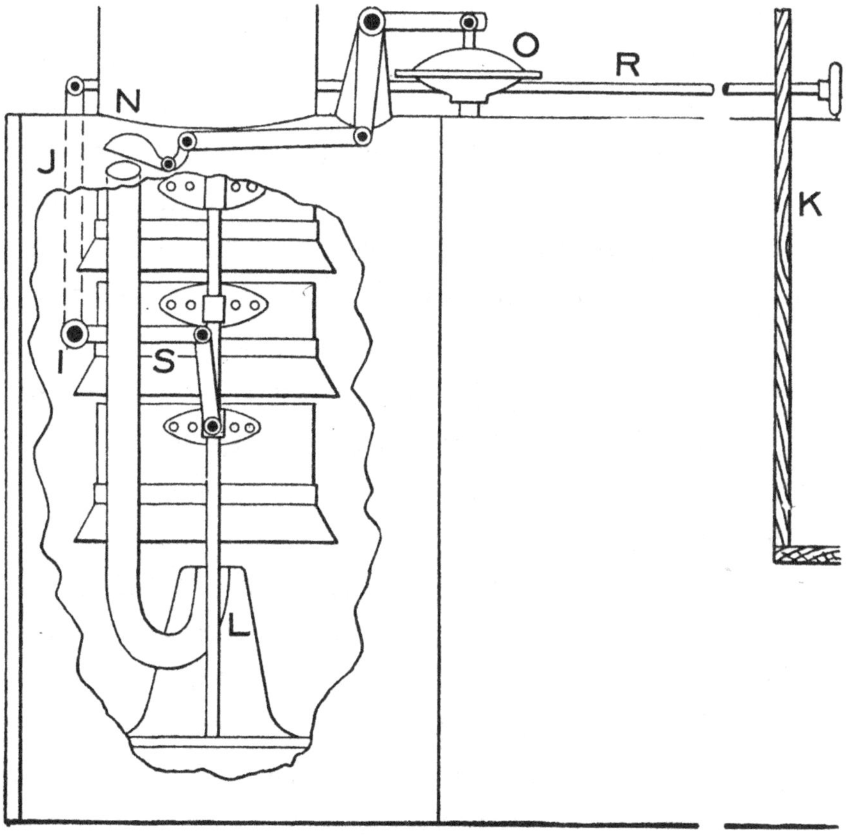

FIGURE 48

which runs to the cab R. By pulling or pushing the rod K the petticoats are raised and lowered, thus increasing and decreasing the distance from the exhaust nozzle L, thereby increasing or diminishing the draft. The air tubes M M turn up alongside the exhaust nozzle L, and are opened and closed by valves N N on the outside of the boiler. The valves are operated by a pressure regulator O, so adjusted that they are opened by the steam when it passes a given pressure. This operates on the crank P and connecting rod Q to open the valve, thus admitting air to the smoke-box and decreasing the amount drawn through the tubes and decreasing the consumption of the coal, and obtaining the full benefit for all fuel consumed without letting the cold air in onto the hot iron. By this means we have the combustion automatically regulated, also obtaining the greatest amount of heat from the fuel consumed.

The petticoat or draft pipe is a very important factor in the regulation of the draft in a locomotive, so as to have the fire burn equally in all parts of the fire-box. Sometimes the fire is inclined to burn the strongest at the back end of the fire-box. This is caused by the draught pipe being set too low. On the other hand, if the fire burns the strongest at the front it shows that the petticoat pipe is too high and it should be lowered.

The exhaust nozzles become at times coated with a hard, gummy substance on the inside, thus decreasing their area, and the result of this is that the fire is torn and cut to pieces on account of the too strong draft The remedy for this is to ream out the nozzles by means of a reamer having a long handle whereby it can be introduced through the stack.

Another device for regulating the draft is used in the extended smoke-box. This is a diaphragm placed at an angle of 20 degrees usually, although some high authorities advocate placing it at 30 degrees. The gases impinge against the diaphragm, and are thus impeded in their passage to the stack, the flow being regulated by means of a diaphragm damper.

One very important requisite for obtaining good combustion and an even burning fire in a locomotive is that the exhaust should fill the stack, and not go through it like a shot out of a cannon, chopping the fire and carrying with it green coal and a large volume of gases that are unconsumed. Close observation and careful work are needed to guard against the great waste of fuel caused by an incorrect adjustment of the various factors contained within the smoke-box or front end.

The raising of the apron or damper on the diaphragm will give more draft through the top flues and cause the fire to burn more brightly at the back of the fire-box, and to lower the apron causes a stronger draft through the lower tubes and a consequent harder burning of the fire at the front. In experimenting along this line, a change of a quarter of an inch at a time is sufficient until the proper position for the apron is found.

The following timely observations on the locomotive front end are from the pen of Mr. K. P. Alexander, master mechanic of the Ft. S. and W. R. R., and were published in the May, June and July, 1905, issues of Railway and Locomotive Engineering. By permission of Mr. Angus Sinclair, editor of that valuable journal, the article is here inserted.

"For much of the data used in this paper I take pleasure in acknowledging indebtedness to Prof. W. F. M. Goss of Purdue University, and committee reports of the American Railway Master Mechanics' Association for blue prints, reports, personal information, etc.

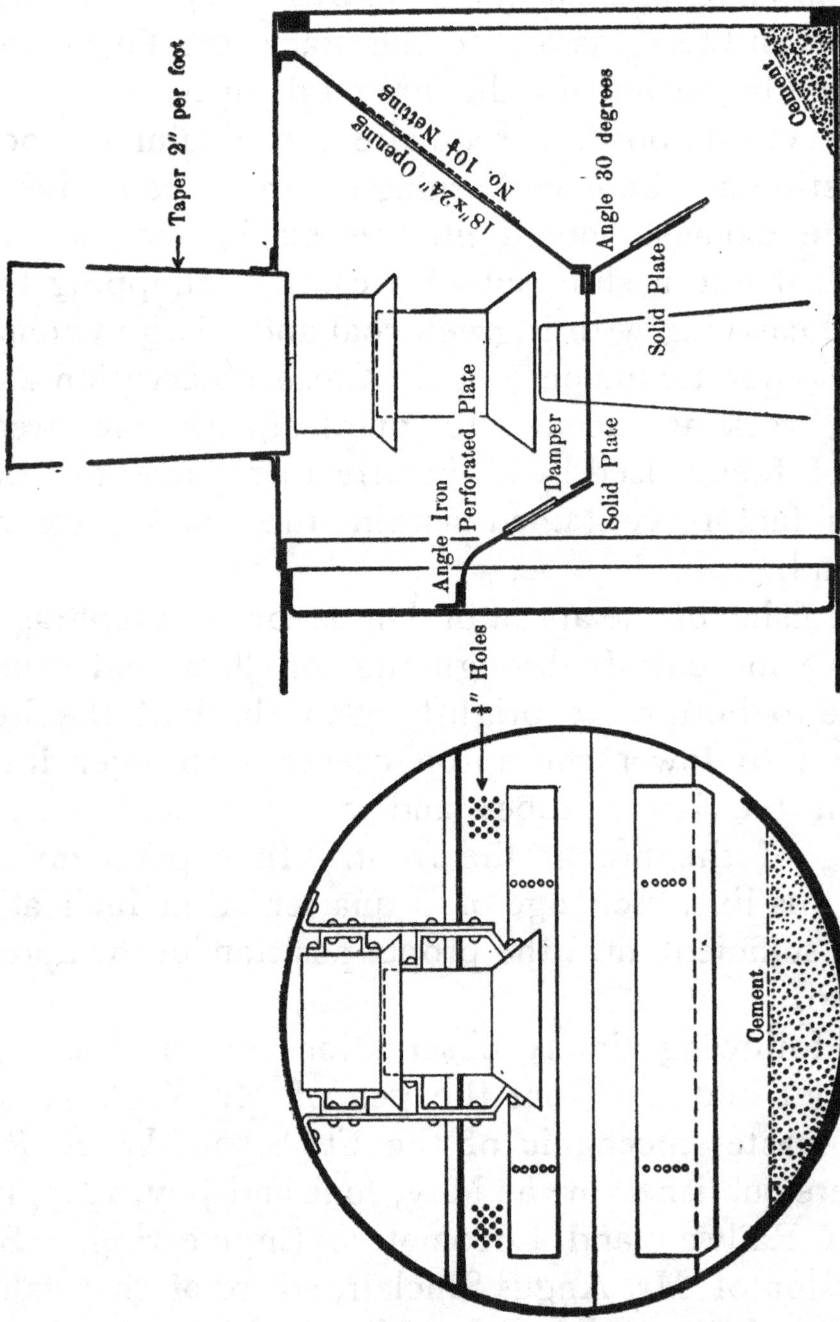

THE "FRONT END"

Without presuming to give as much detailed information, this article is intended to more completely embrace the known facts relating to the several parts of the front end than any single report that has appeared.

The "front end" includes the diaphragm, the exhaust nozzle, the exhaust stand, the stack, the petticoat pipe, and the netting. These, with the exhaust jet, constitute an apparatus designed to produce the maximum amount of draft through the fire with the minimum of back pressure in the cylinders. The efficiency of the front end is therefore the greatest possible ratio of draft to back pressure.

The Diaphragm. The total draft is said to have three approximately equal factors of resistance to overcome: the diaphragm and netting, the flues, and the fire, grates and ash pan. As the diaphragm (or baffle plate) absorbs about one-third of the energy of the exhaust jet, the net efficiency of the front end is evidently increased as the angle of the diaphragm is changed from the usual angle of 20 degrees toward a more horizontal position, or to an angle of probably 25 degrees. Within certain limitations, the front end is also increased in efficiency by enlarging the area of opening under the diaphragm damper. Indeed, it is said that there are foreign railroads that have in some manner successfully dispensed with the diaphragm and yet secured equalization of draft over the entire fire-box.

The opening under the diaphragm damper must, however, be of such width horizontally as will allow of an area of opening equal to the total cross-sectional area of the opening through the flues, and at the same time be sufficiently contracted to retard the flow of gases from the fire-box long enough to consume as great a per cent of the gases as affecting conditions will permit. It must also be sufficiently contracted, in self-cleaning front ends, to obtain enough velocity to keep the front clean of sparks. The diaphragm damper, or movable deflecting plate, must be set at such height as, with a given angle of the diaphragm, will produce a slightly stronger draft at the back than at the front end of the fire-box. The area of opening under the diaphragm should be greater for slow-burning than foı free-burning coal, as, by diminishing the non-effective work that must be performed by the exhaust jet, the nozzle may be enlarged or a greater per cent of its effective energy may be utilized in producing draft through the fire.

Draft through the fire in the back end of the fire-box is increased by increasing the angle of the diaphragm, or by raising the diaphragm damper, also by about four horizontal rows of holes punched in the upper end of the top section of the diaphragm. As the amount of draft is proportional to the weight of steam exhausted per unit of time, it is believed that differences in grate area do not materially change the volume of gases passing under the diaphragm. Contracted opening under the diaphragm, or through the grates, probably results in slight cylinder back pressure. When the area of opening under the diaphragm is enlarged by raising the diaphragm damper beyond a certain limit, the angle of the diaphragm must be decreased. The effect of wings projecting at each end of and below the diaphragm is to decrease the draft at the side sheets and to concentrate it along the center of the fire-box. The length of the horizontal part of the diaphragm (the distance between the upper and lower sections) does not affect the draft in either end of the fire-box.

The most efficient diaphragm should give the following: most rapid rate of combustion, slightly stronger draft at the back than at the front end of the fire-box, sufficiently baffle the flow of gases so as to result in the most complete combustion that the affecting conditions will allow, with minimum resistance to the exhaust jet.

Such a diaphragm should have an angle of about 25, instead of 20 degrees. Just below the arc of a 5-in. radius bend in the top end of the upper section should be punched about four horizontal rows of 3/8-in. holes with 3/8-in. centers, extending across the upper section a distance equal to the distance between the steam pipe centers at that height. An adjustable damper should be applied, to regulate the area of opening through the holes, in order that the proper degree of draft may be obtained in the back end of the fire-box. On the back side of both sections of the diaphragm should be bolted perforated steel plate with $\frac{3}{16} \times 1\frac{1}{2}$ in. mesh, set with slots vertical.

The object in increasing the angle of the diaphragm from 20 degrees (the usual angle) to 25 degrees, is to diminish the resistance to the exhaust jet. But, with such a change in angle, the four rows of 3/8-in holes in the upper section are necessary in order to increase the draft in the back end of the fire-box as much as the change of angle increased it in the front end of the fire-box. The perforated steel plate bolted to the back side of the diaphragm very materially assists in breaking up the cinders as they strike it at an angle, thus considerably increasing their facility in passing through the netting and decreasing the liability of starting fires. This is equivalent to increasing the netting area or enlarging the opening of the mesh, and therefore lessens the total amonnt of work that must be performed by the exhaust jet. The horizontal plate of the diaphragm should always, regardless of the height of the exhaust stand, for self-cleaning front ends, be located just under the top flange of the exhaust stand. In order to get in sufficient netting for free steaming this plate should never be set higher than 2 in. below the center line of the smoke-box, nor more than 6 in. below the top of the nozzle.

The Exhaust Jet and Nozzle. The most accurate, reliable and comprehensive data on the form, density and efficency of the exhaust jet, is contained in the 1866 report of a committee of the American Railway Master Mechanics' Association, under the chairmanship of Robert Quayle of the Chicago and Northwestern Ry. The matter in this paper referring to the exhaust jet, especially the measurements of vacuum and pressure in stack and front end, is largely based on that report.

The cross-sectional form of the exhaust jet is influenced by the form and dimensions of the channel surrounding it, even though not in actual contact. It is supposed that, in the stack, the vacuum around the column of the exhaust tends to compact it and thus prevent contact with the stack until it reaches nearly to the top of the stack. Whether this is true or not, personal experiments indicate that when the surrounding channel is within a certain distance of a column of steam issuing from a taper nozzle, the jet is apparently attracted to and comes in actual contact with the enclosing channel. Accurate tests made by the Master Mechanics' Association Committee show beyond question that the exhaust jet does not, and preferably should not, fill the stack at or near its base, but that it comes in contact with the stack only quite near the top. The foregoing facts should be remembered in connection with calculating the diameter of petticoat pipes. The plan of the angle of the exhaust jet is not like an inverted frustrum with sides of straight lines, as is commonly supposed. Its form, between the nozzle and its point of contact with the stack, is represented by two slightly concave curved lines. It is in actual contact with the stack only about 10 or 12 in.

Vacuum gauges (measured in inches of water) show that the vacuum between the wall of the stack and the column of the exhaust jet, at a point one-third of the length of the stack from its top, is 1.50, midway of its length it is 2.52, and at about 17 in. from its base it is 3.61. At a point midway between the smoke-box circumference and the nozzle, on a line with the center of the arch, the vacuum is 2.54.

The pressures in the center of the exhaust jet are, at about 12 in. above the nozzle, 59.3; 24 in. above the nozzle, 44.6, and about 6 in. below the top of the arch, 28.5. The gauge also showed that the pressure diminished rapidly as it was moved from the center toward the circumference of the jet, varying in velocity from 576 to 292 ft. per second. Increasing the number of pounds of steam exhausted per unit of time, or increasing the boiler pressure, increases the velocity and diminishes the spread of the jet, resulting in increasing the vacuum.

The direction of the gases in every part of the smoke-box and stack is from the nozzle tip up toward the exhaust jet, and not directly toward the stack. The smoke-box gases and sparks are slightly enfolded within, but largely entrained by the exhaust jet. The induced action of the jet is greatest and the intermixing or enfolding action least, at the nozzle. It is believed that as the mixing action is increased the induced action is diminished, with no resulting gain, and that therefore the more compact the jet the higher will be its net efficiency.

It is claimed that the efficiency of the jet is unchanged, providing the weight of steam exhausted per unit of time is equal, whether the engine is working at long cut-off with heavy impulses of the exhaust at long intervals, or working at short cut-off with quicker or lighter impulses at shorter intervals. The nozzle diameter should be as great as affecting conditions will permit.

Increasing the rate of combustion by undue contraction of the nozzle or grate area results in considerable decrease in evaporation per pound of coal. This is due to back pressure in the cylinders and to excessive spark losses and incomplete combustion of the gases in the fire-box. Increasing the rate of combustion per square foot of grate surface per hour from 61.4 to 240.8 lbs., decreased the evaporative efficiency 19.2 per cent and increased the pounds of sparks per hour from 46 to 160 lbs.

There is doubt as to whether a splitter or bridge in the nozzle is of any benefit under any possible conditions. However, apparently good results have been obtained by enlarging a nozzle equal to the cross-sectional area of a ¼-in. or ⅜-in. splitter, when such splitter was placed in the top of the nozzle at right angle to the partition in the exhaust stand. Any possible advantage of such a bridge would be its effectiveness in overcoming the form of the exhaust (in an exaggerated form represented by the shape of a figure 8) due to the action of the exhaust jet in exhausting somewhat from side to side instead of exactly vertical.

this being due to the deflecting influence of the exhaust stand partition and the inner angle of the nozzle.

The most efficient form of exhaust nozzle is the single one, with its interior in the form of a frustrum of a cone, ending at the top end with a parallel cylinder 2 in. long. The distance from the nozzle to choke of 14-in. stack 52 in. long, on a 58-in. front end, should not exceed 50 in. or be less than 40 in., for maximum efficiency. The distance from nozzle to top of smoke arch with a 14-in. straight stack 52 in. long should not be less than 22 in. nor greater than 38 in. The distance from nozzle to top of arch with a 16-in. straight stack 52 in. long should not be less than 28 in. nor greater than 38 in. The distance between nozzle and choke of stack should be slightly increased for the highest steam pressure.

The Exhaust Stand. The cross-sectional area of choke in each side of exhaust stands (when choked at all) should at least equal the area of the largest nozzle that may be applied. Bulged, or pear-shaped, stands are objectionable on account of interfering with the free passage of the gases from under the diaphragm damper. Stands should be not less than 19 in. high. They should have a partition in them to prevent the exhaust from one side effecting back pressure in the other side of the engine, but such partition should not be less than 8 in. nor more than 12 in. high, and it should not extend a greater height than to a point 10 in. from the top of the stand.

The Stack. For a 54-in. front end, the highest efficiency is obtained by a tapered stack, tapered 2 in. per foot, with its smallest diameter a distance of 17½ in. from its base. The greater the height of stack the greater will be its efficiency. Tapered stacks, whether long or short, should equal in diameter at inside of choke one-fourth of the diameter of the arch. The diameter of the stack should be diminished as the nozzle is raised.

Professor Goss gives the following formula for determining correct nozzle heights. H equals height of stack, h equals distance in inches between center line of boiler and nozzle, d equals diameter of choke of stack, and D equals diameter of front end.

Formula for Tapered Stacks. When nozzle is below center line of boiler: $d = .25\ D + .16\ h$. When nozzle is above center line of boiler: $d = .25\ D - .16\ h$. When nozzle is on center line of boiler: $d = .25\ D$.

Formula for Straight Stacks. When nozzle is below center line of boiler: $d = (.246 + .00123\ H)\ D + .19\ h$. When nozzle is above center line of boiler: $d = (.246 + .00123\ H)\ D - .19\ h$.

The Petticoat Pipe. As a means of increasing the induced action of the exhaust jet, rather than as a means of equalizing front and back the draft on the fire, double petticoat (or draft) pipes add to the efficiency of the front end. When the distance between the nozzle and the choke of the stack (the top of the arch, with a straight stack) is not great enough to make a double pipe practicable, a single pipe is beneficial. The efficiency of the draft pipe is mainly due to its forming a longer orifice through which the exhaust must pass, thereby augmenting the induced action of the exhaust jet by solidifying it, it not being essential or desirable that the jet come in actual contact with the draft pipe. In fact, the pipe should be so large that the jet will not touch it.

In a 58-in. front end the best results were obtained with a 14-in. choke stack, choke 12 in. above top of arch, nozzle 45 in. from choke, with a double petticoat pipe. The highest net efficiency was when the bottom end was set even with (but none below) the top of the nozzle. The top end of the upper section was set 13½ in. below the choke of the stack. The total distance from nozzle to top of upper section, in this position, was 28½ in. The smoke-box vacuum decreased as the distance was lengthened to 31 in., and the back pressure in the cylinders increased as the distance was shortened from 29 to 28 in. The double petticoat pipe used in above test was of following dimensions: lower section, 10 in. diameter by 11 in. long; upper section, 13 in. diameter by 10 in. long. The flare on lower section was 7 in. high by 17¼ in. diameter at bottom; flare on upper section was 2 in. high by 15 in. diameter at bottom end.

The Netting. No data is on record of the amount of resistance to the exhaust jet due to the front end netting, or perforated steel plate. The total area of netting should be as great, and its mesh as large, as conditions will, with safety, permit, as the open area is considerably reduced at each impulse of the exhaust by sparks in process of being broken up sufficiently small to pass through. As the direction of the sparks in the smoke box is from every point toward the column of the exhaust jet, instead of directly toward the stack, the netting should be set so that, as nearly as may be, the sparks will strike it at right angle to its face.

Although some railroads use coarser and some finer mesh, it is probable that the most preferable is netting with 2½ x 2½ mesh No. 10½ double crimped steel wire, or $\frac{3}{16}$ x 1½ in. perforated steel plate, with the plate set so that the slots run vertically instead of horizontally. The chief objection to the perforated steel plate is that it necessarily contains less open area in proportion to its closed area than netting. A point in its favor, however, is that sparks cannot as easily wedge in the perforations as in the mesh of the netting."

Questions

122. What are the main factors in the transmission of the steam from the boiler to the cylinders?

123. Describe the steam dome.

124. What is the object in placing a dome on a locomotive boiler?

125. How is the steam conducted from the boiler to the cylinders?

126. Where are the steam pipes located?

127. Where is the throttle usually located?

128. Describe the old style of throttle.

129. What was the objection to such a throttle?

130. Describe in general terms the modern improved throttle.

131. Why is the lower disk smaller in diameter than the upper one?

132. What kind of a joint is used for connecting the steam pipes to the dry pipe within the smoke-box?

133. Describe a ball joint.

134. What other kind of joint is often used for making these connections?

135. What are the exhaust nozzles?

136. Why are rings or bushings usually fitted in the outlets of exhaust nozzles?

137. If the exhaust orifice is too large, what is the result?

138. What is the effect upon the fire if the exhaust outlet is too small?

139. What is an adjustable exhaust nozzle?

140. What is the function of the petticoat, or draft pipe?

141. If the draft pipe is set too low, how is the fire affected?

142. If the fire burns too strong at the front, what should be done with the draft pipe?

143. Why is it necessary to ream out the exhaust nozzles at times?

144. What other device for regulating the draft is placed in the extended smoke-box?

145. At what angle is the diaphragm usually placed?

146. What effect does this have upon the gases on their way to the stack?

147. How is regulation of the draft accomplished with the diaphragm?

148. What is a very important requisite for good combustion in a locomotive fire-box?

149. How does the raising of the apron or damper of the diaphragm affect the burning of the fire?

150. How will the fire be affected if the apron is lowered?

151. How much should the apron be moved up or down at a time, when making adjustments for draft?

BIG FOUR

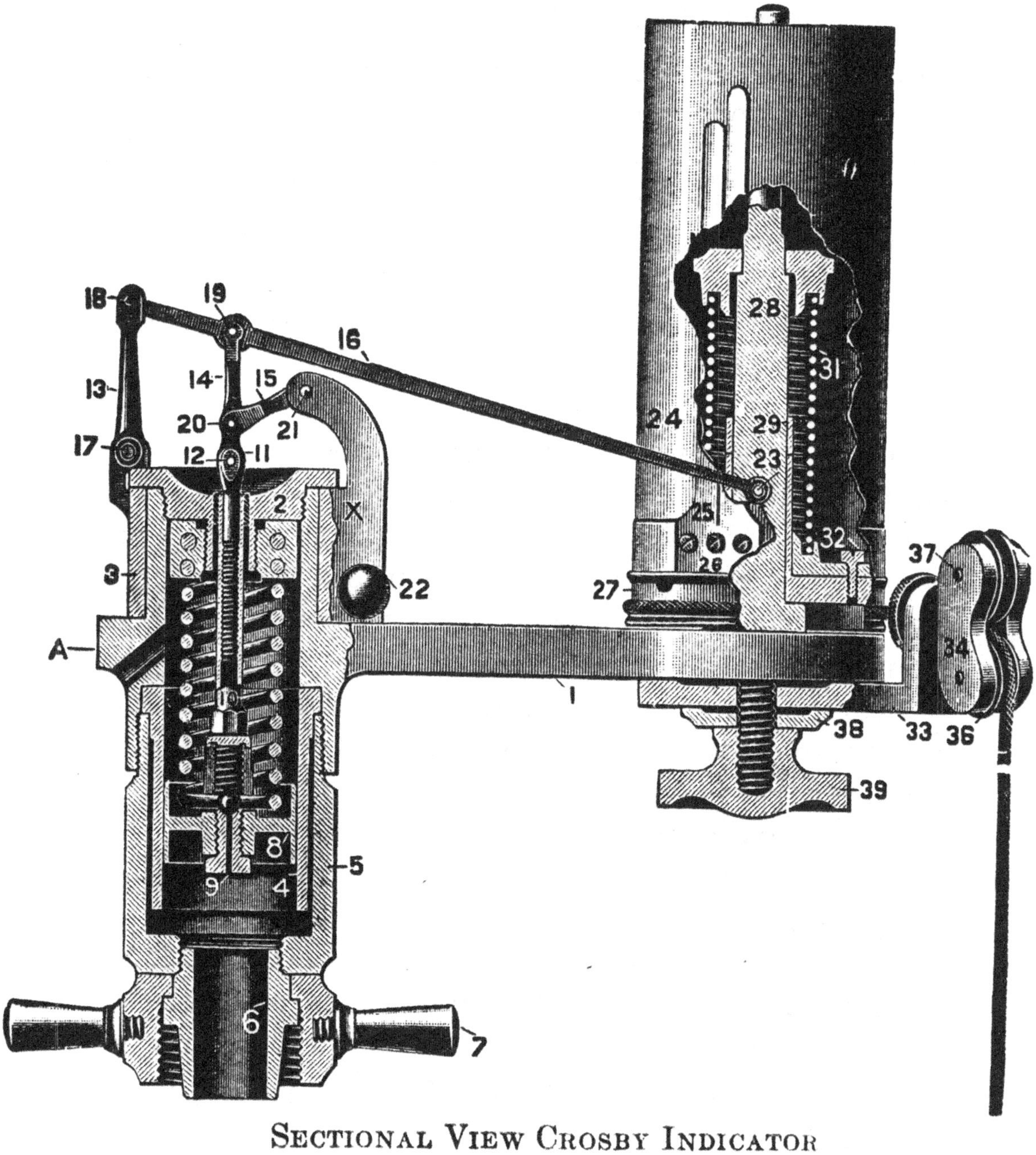

Sectional View Crosby Indicator

CHAPTER IV

VALVES AND VALVE GEAR

In a certain "catechism" of the locomotive the following question and answer appear: "Q. What is a locomotive? A. A locomotive is a boiler and two or more engines mounted on wheels." This answer, while not very definite, is certainly "short and to the point."

Two types of locomotive engines are in use, viz., simple and compound.

A simple engine, whether stationary or locomotive, is an engine in which the steam is made to do work in but one cylinder, after which it is exhausted into the atmosphere, or, as is the case with many stationary engines, the exhaust steam passes into a condenser in which a vacuum is maintained, and the steam is there condensed.

A compound engine is an engine in which the steam is made to do work in two or more cylinders before it is allowed to escape into the atmosphere or condenser. The expansive properties of the steam are thus utilized in a much higher degree than with the simple engine, and great economy in fuel is the result.

As an entire chapter is devoted to compound locomotives, the subject will not be enlarged upon at this stage, but the attention of the student will now be directed to a study of the valves, valve gear, etc., of a simple engine.

In Fig. 31, Chapter III, is given a sectional view through the smoke-box, saddle plate and cylinder castings of a simple engine, having a flat or D slide valve. Figs 49 and 50 show respectively a front end view of the cylinder, valve chest and saddle plate castings, and a section through the same parts showing the steam and exhaust passages. These castings require to be of the best grade of iron, neither too soft nor too hard.

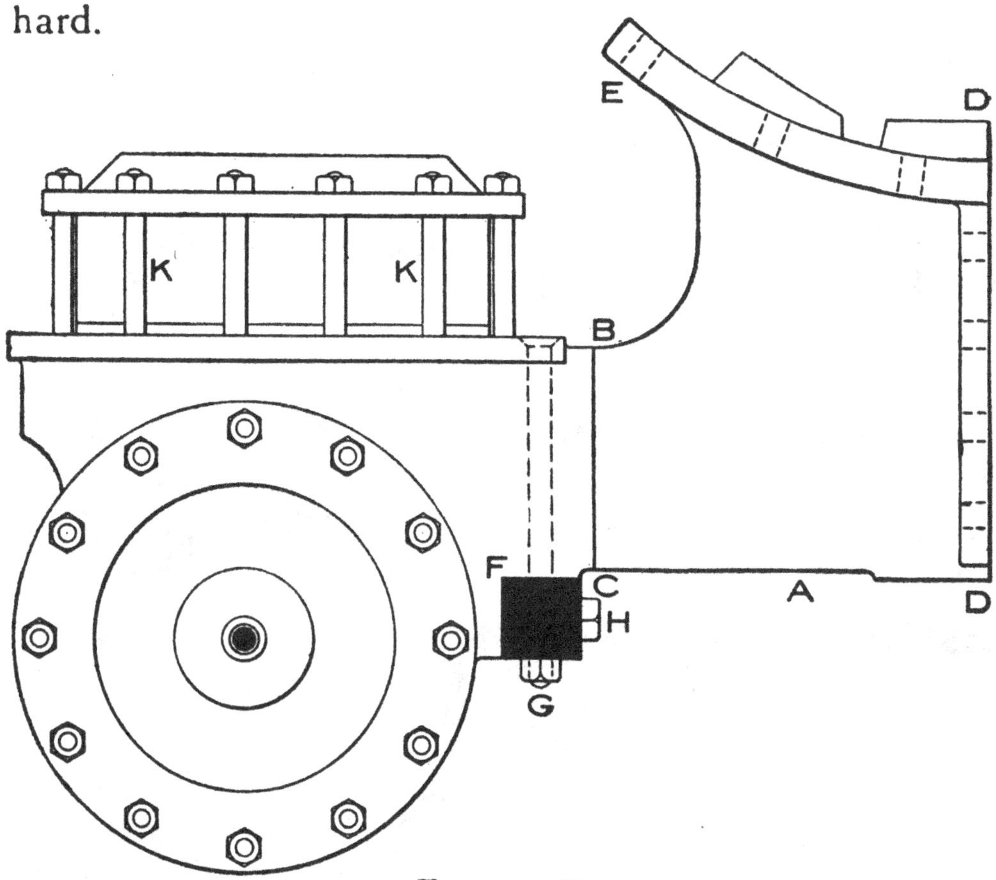

FIGURE 49

Cylinders and valve seats are generally cast together with the bed plates or bed castings A A, Figs. 49 and 50. Sometimes the bed castings are made separate from the cylinders and in one piece, and the cylinders are then bolted to it about at the line B C, Fig. 49. The usual practice, however, is to cast one-half of the bed casting with each cylinder and then bolt the two halves together at the line D D, this being the center line of the engine.

The bed castings are secured to the smoke-box by bolts through the flanges E E, and the cylinders are bolted to the frame F F by bolts G and H, Figs. 49 and 50. A structure is thus formed that is able to withstand the tremendous strains to which it will be subject.

By reference to Fig. 50 it will be seen that there are

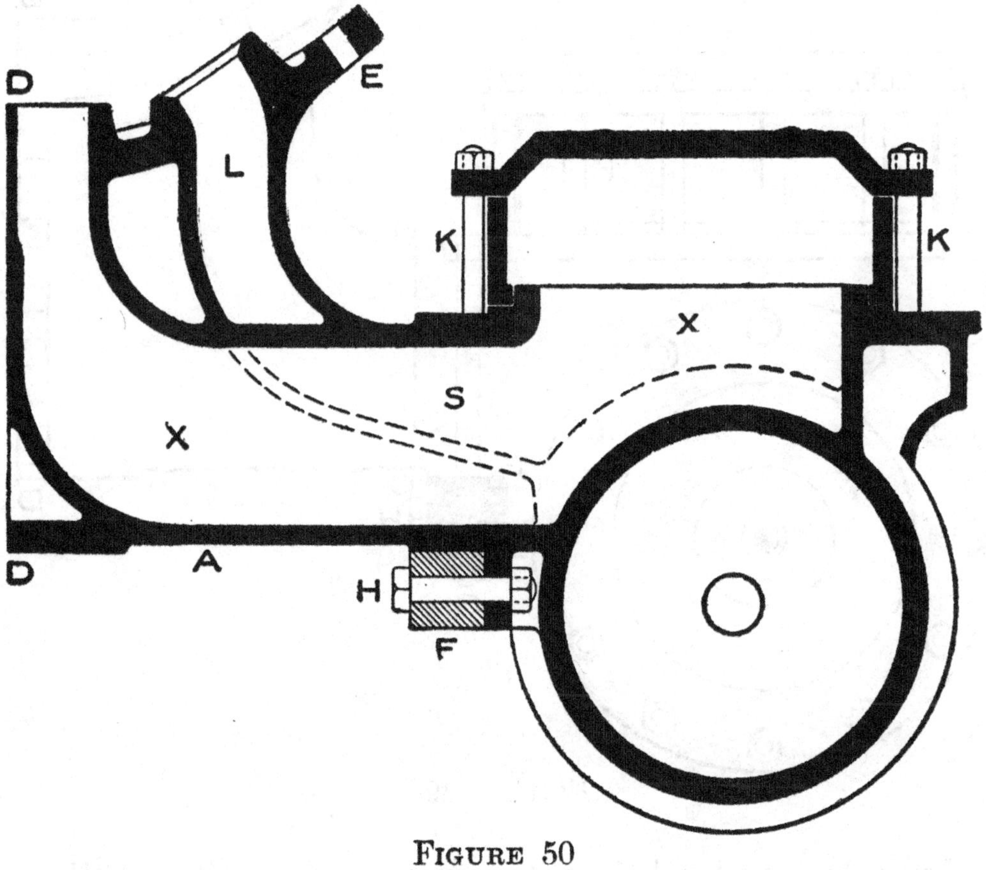

FIGURE 50

two passages in the bed casting leading to the cylinder. The one L S is the live steam passage, and to this is connected the steam pipe 84, Fig. 31, Chapter III, and the other one *x x* is the exhaust passage leading from the cylinder to the smoke-box where the exhaust nozzle is attached.

Fig. 51 gives a longitudinal sectional elevation of the cylinder, valve chest, valve, etc., showing the steam passages more clearly. Fig. 52 is a plan of the cylinder and guides and shows the valve seat with its ports, the steam chest cover and valve being removed. The steam passage A on approaching the steam chest is divided into two branches, which terminate in open-

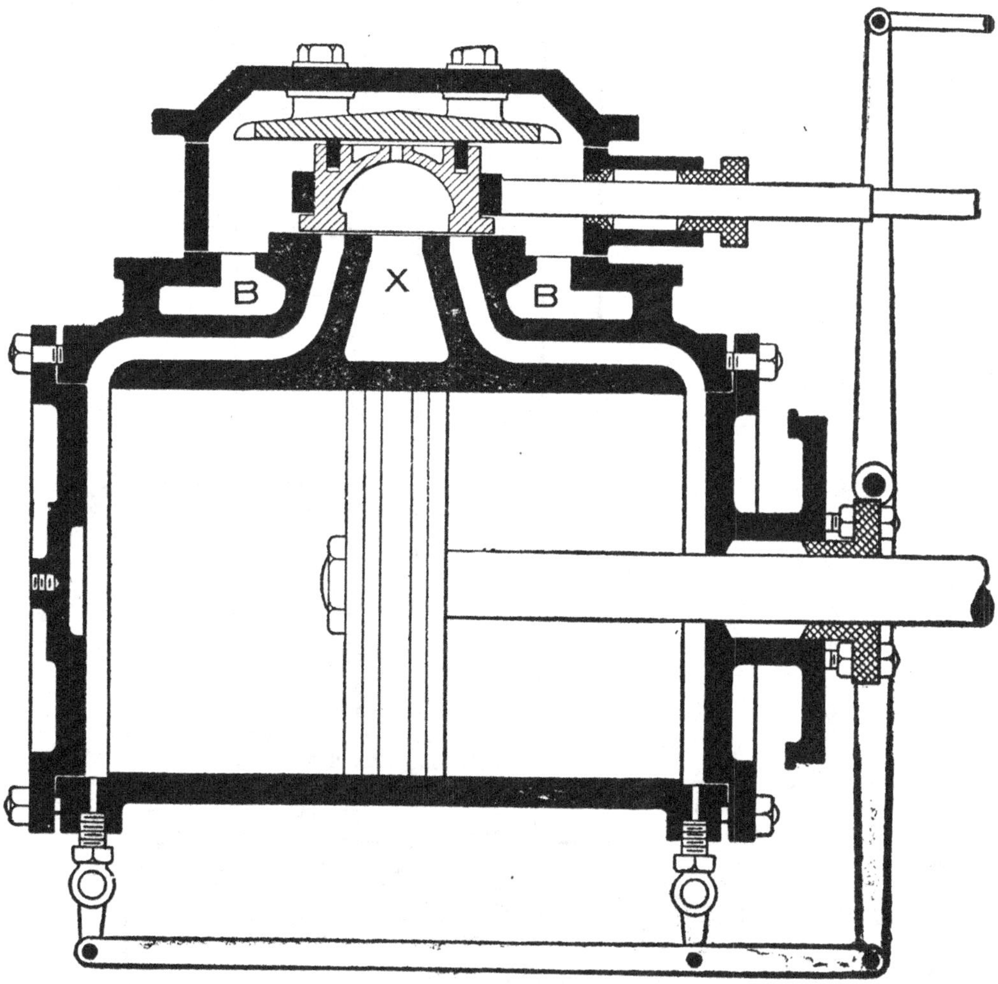

FIGURE 51

ings B B at each end of the valve seat (see Figs. 51 and 52). This causes the steam to be delivered at both ends of the steam chest and on top of the slide valve, which covers the ports P P and the exhaust port *x x* when in its central position (see Fig. 51).

The steam chest is a square cast iron box open at

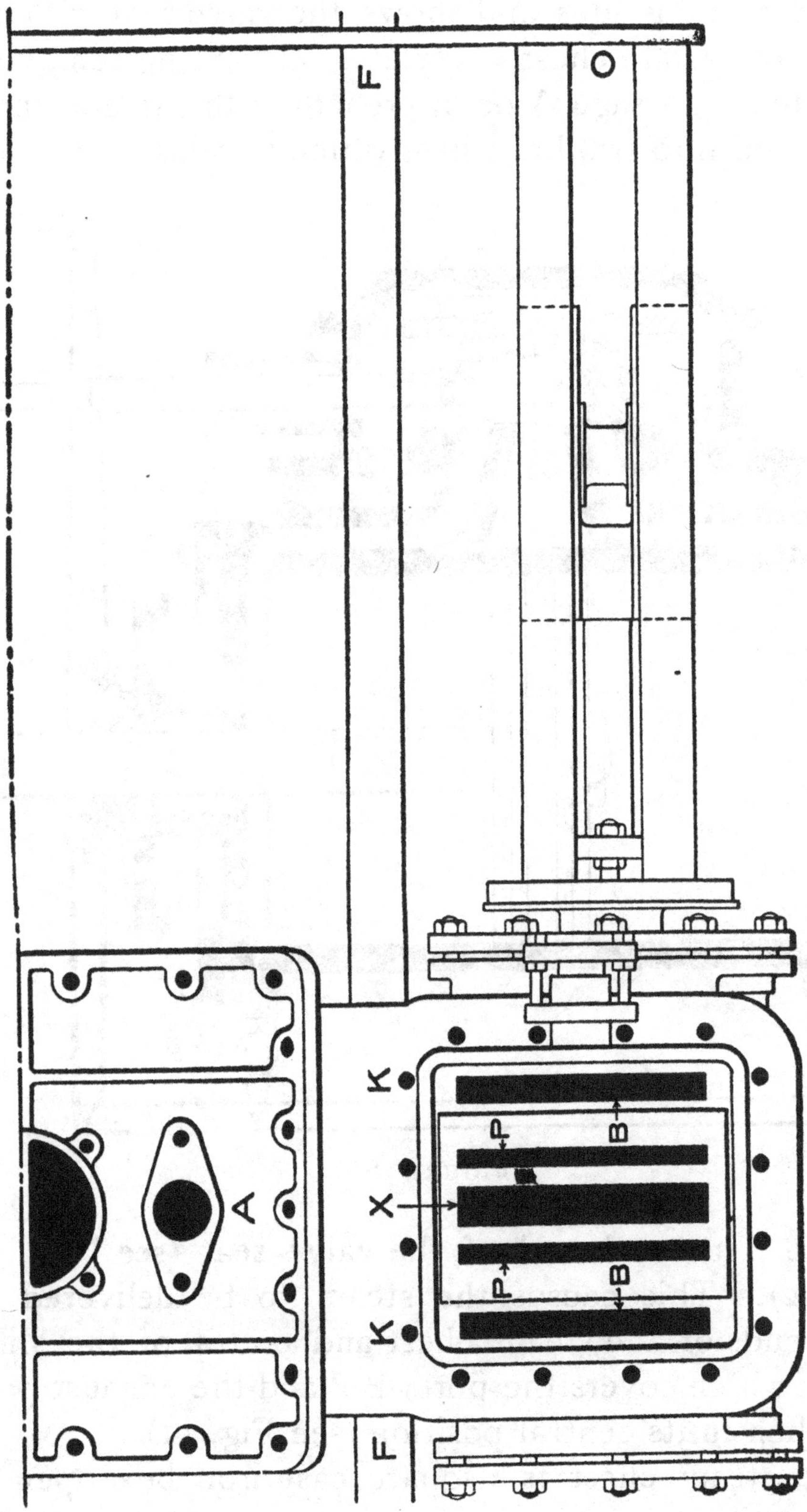

FIGURE 52

top and bottom and resting upon the top of the cylinder casting. The steam chest cover rests on top of this box and the whole is held down upon the cylinder by strong bolts K K, Fig. 49, forming a steam-tight joint, and within this valve chest the slide valve performs its important work.

The invention of the D slide valve in its present form is the result of the investigations of Murdoch, who was an assistant to James Watt, the man who contributed more than all others towards the development of the steam engine and its practical application. No young man who aspires to become a locomotive engineer should rest satisfied until he has obtained a thorough knowledge of the construction, operation, and adjustment of the slide valve. Many have been the efforts made to displace it with other types of valves, and while no doubt in stationary work other forms of valves may be better adapted to the conditions, yet the slide valve in some form or another still holds its own with the locomotive.

The functions that a slide valve must perform in order that the engine may do efficient work are five in number, and they are as follows: First, it must admit steam into one end only of the cylinder at the same time. Second, it must cover the steam ports so as not to permit the passage of live steam through both steam ports at the same time. Third, it must allow the steam to escape from one end of the cylinder before it is admitted at the other end, so as to give the steam that is to be exhausted time to escape before the piston commences the return stroke. Fourth, it must not permit live steam to enter the exhaust port direct from the steam chest. Fifth, it must close each steam port on the steam side before it is opened on the exhaust side; this is for the purpose of utilizing the expansive force of the steam.

Figs. 53 to 58, inclusive, show the general construction of the D slide valve and illustrate the various positions assumed by it during one stroke. Fig. 53

FIGURE 53

shows the valve in its central position and explains the meaning of the word lap. Outside lap, often referred to as steam lap, is the distance that the edge of the valve overlaps the steam ports when it stands central or at mid-travel, and is that portion of the

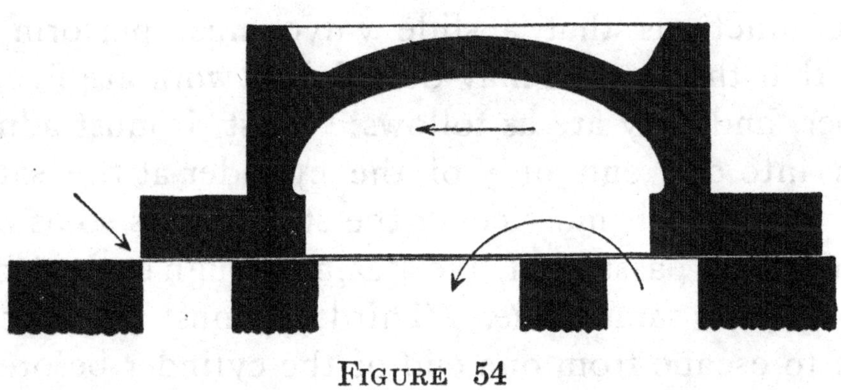

FIGURE 54

valve marked L and indicated by the distance between the lines O P. Inside lap, frequently referred to as exhaust lap, is that portion of the valve that overlaps the two bridges of the valve seat when the valve stands central, and is shown at A and A, Fig. 59.

Inside clearance, sometimes called exhaust lead, is the space between the inside edges of the exhaust arch of the valve and the bridges when the valve stands central. It means just the reverse of inside lap; that is, the distance between the inside edges of the exhaust

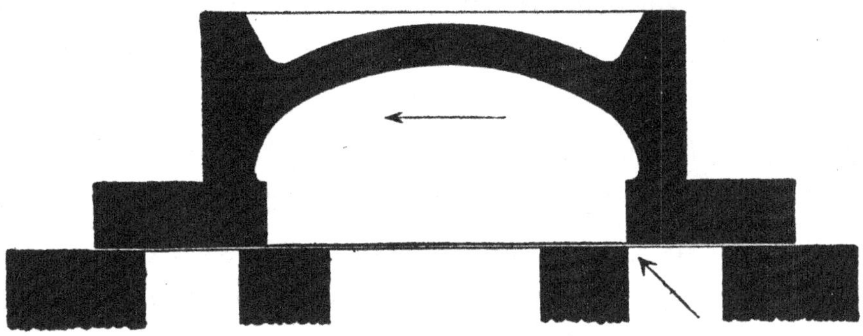

FIGURE 55

arch is slightly greater than the distance between the inside edges of the steam ports, so that it does not entirely cover them when in its central position. Inside clearance is shown at B and B, Fig. 60. The purpose of inside lap is to delay the release of the

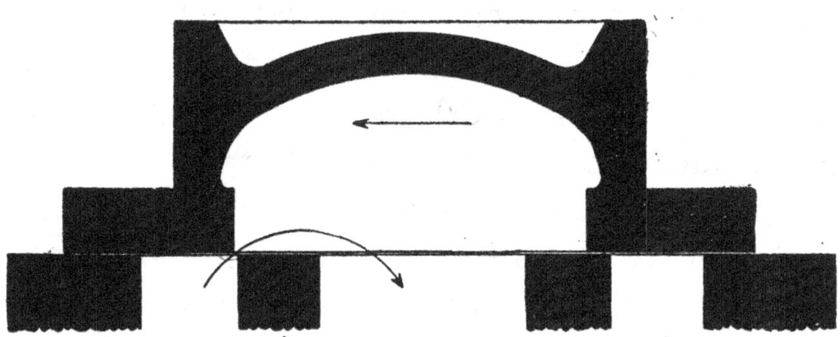

FIGURE 56

steam and to hasten compression. The amount of inside lap is small, seldom exceeding ⅛ of an inch, and for fast passenger engines it is better to have none, as it causes the engine to be quicker. The purpose of inside clearance is exactly the opposite to that of inside lap. It hastens release and delays compression. It rarely exceeds $\frac{1}{64}$ of an inch, and is only used on very fast running engines. Good judgment and great experience are required in order to determine the proper amount of inside clearance and upon what classes of locomotives it should be used.

In locomotives for ordinary service the valves have no inside clearance. Cut-off, Fig. 54, refers to the closing of the steam port by the valve, thereby cutting off the flow of live steam to the cylinder before the piston has completed its stroke. Compression refers to the early closure of the passage between the cylinder and the exhaust port. This point is reached when the inside or exhaust edge of the valve has closed the steam port, as shown in Fig. 55, wherein the valve is assumed to be traveling in the direction indicated by the arrow. A small portion of the steam is thus retained in the cylinder to be compressed by the advancing piston, which thus meets with a slight cushion at the end of its stroke, and all shock and jar is thus prevented. Release occurs when the exhaust edge of the valve opens the steam port and allows the steam that has completed its work in the cylinder to escape into the exhaust port, as shown in Fig. 56.

Lead, otherwise called steam lead, is the amount of opening given to the steam port by the valve for the admission of live steam to the cylinder when the piston is at the commencement of its return stroke. The lead is indicated by the letter A in Fig. 57.

Travel is the distance through which the valve travels, otherwise its stroke. Over travel is the distance the steam edge of the valve travels after the steam port is wide open, indicated by distance between lines O and T, Fig. 58.

The objects aimed at in giving a valve outside or

steam lap are: First, that the steam may be cut off before the piston reaches the end of its stroke, and the steam thus enclosed within the cylinder be made to do the work throughout the remainder of the stroke by reason of its expansive properties, and secondly, it

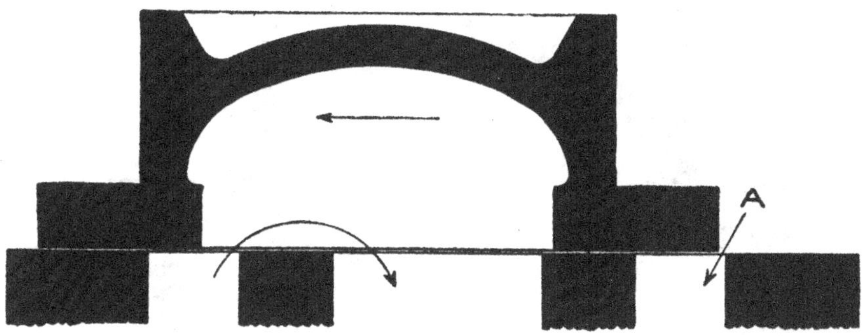

FIGURE 57

causes the exhaust port at one end of the cylinder to be opened before the steam port at the other end is uncovered for the admission of steam. If a valve had no outside lap it would admit steam throughout the whole stroke, or in other words, "follow full stroke."

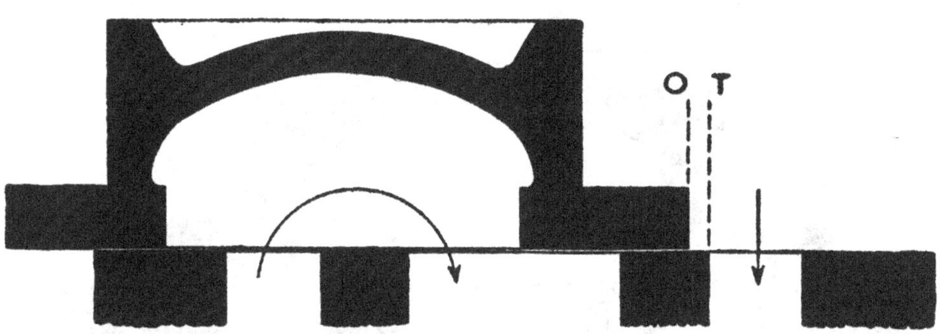

FIGURE 58

Another bad effect of no lap would be a late exhaust, by which is meant that the exhaust would occur at one end of the cylinder at practically the same moment that admission occurred at the other end. This would have a tendency to retard the motion of the piston.

The term clearance, as applied to a locomotive,

means all of the space between the face of the valve and the piston when the latter is at the end of its stroke. Mechanical clearance means the distance between the cylinder head and the piston when at the end of its stroke.

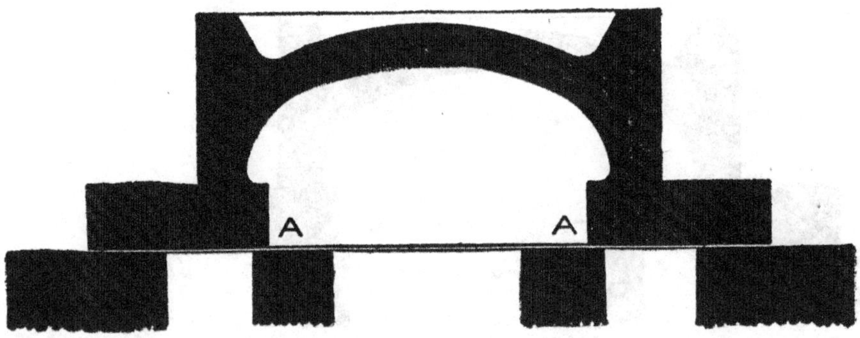

FIGURE 59

The object of giving a valve lead is that the steam port may be opened slightly for the admission of live steam just before the piston reaches the end of its stroke, in order that there may be a cushion of steam to receive the piston and reverse its motion at the end

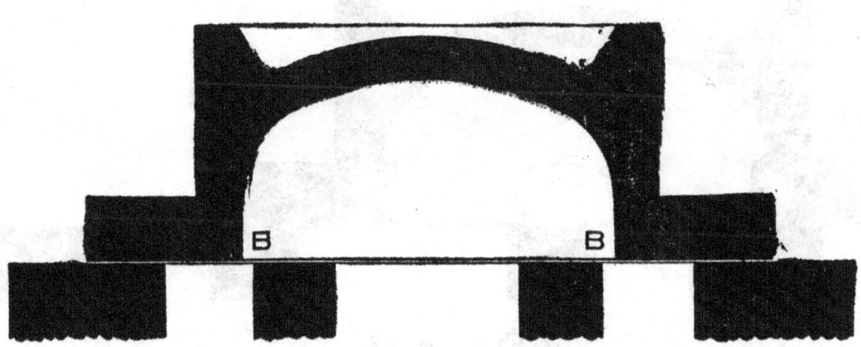

FIGURE 60

of the stroke, thus making the engine quicker in its action. Lead increases on a locomotive as the cut-off is made earlier or shorter, which is done by bringing the reverse lever nearer the center notch of the quadrant. The increased lead is caused by the radius of the link. This will be explained later on, towards the close of this chapter.

The term valve gear, as applied to a locomotive, includes eccentrics, rods, links, rockers, etc., by which the valves are given motion and by which their movements are regulated and controlled. As it is very necessary that a locomotive should be capable of being moved by steam either backward or forward, a reversible valve gear is required. Various devices have been invented for this purpose, but the shifting link motion has, after many years' trial, been found to be the most reliable and best and is to-day the standard in this country.

Fig. 61 shows a general view of the valve gear of one side of a locomotive. The center of the go-ahead eccentric is shown at A, and the center of the back-up eccentric is at B. The eccentric straps are shown connected to the eccentric rods, or blades as they are usually termed, and these in turn are attached to the link, the go-ahead eccentric being connected to the top end of the link and the back-up eccentric to the bottom end. The link saddle S is a plate spanning the center of the link and securely bolted to it. Upon the saddle a pin is formed, to which the lower end of the link hanger is connected, the top end of the hanger being attached to the shorter arm of the tumbling shaft, while the other arm of the tumbling shaft is connected to the reversing rod which extends back to the cab and is connected to the reverse lever.

The link and ends of the eccentric rods connected to it are thus supported and are also free to be moved either up or down by means of the reverse lever. The link block, upon which the link slides freely, is attached by a pin to the lower arm of the rocker shaft,

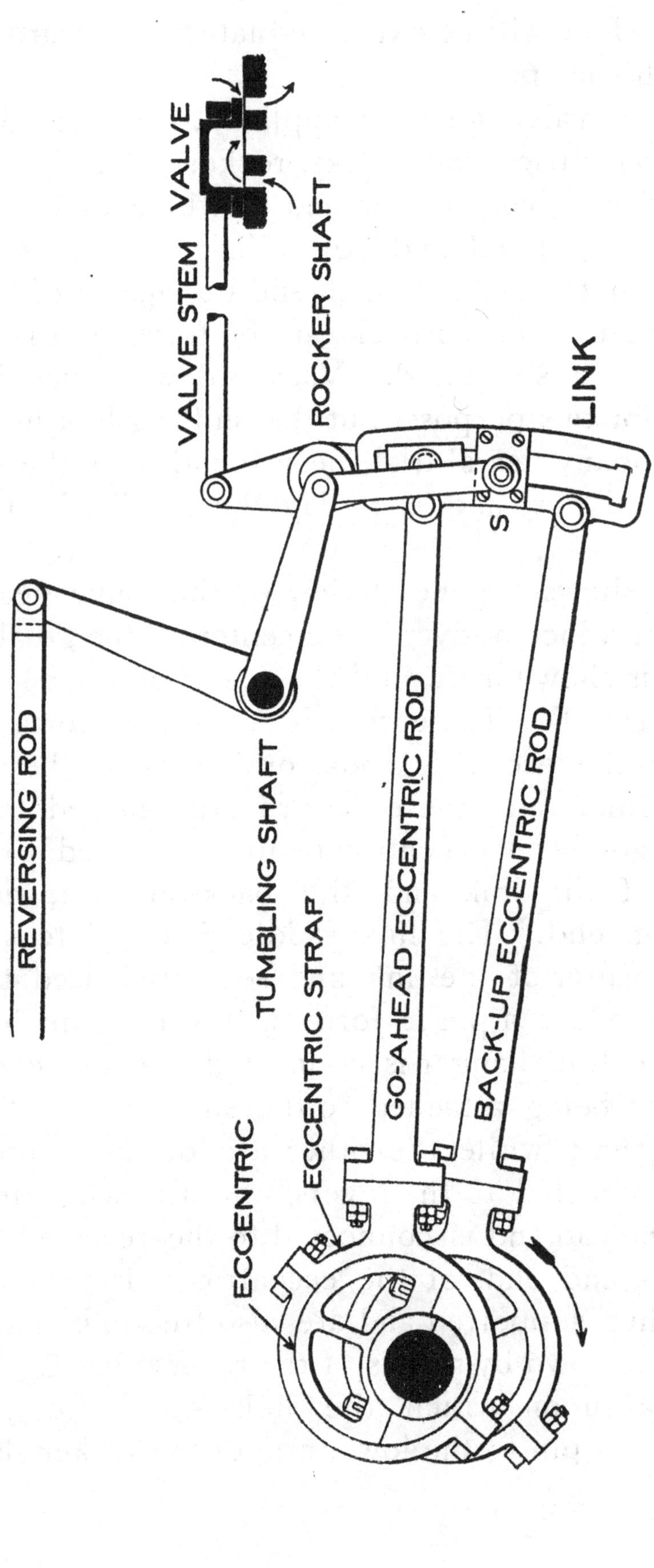
VALVE
VALVE STEM
ROCKER SHAFT
LINK
S
REVERSING ROD
TUMBLING SHAFT
ECCENTRIC STRAP
GO-AHEAD ECCENTRIC ROD
BACK-UP ECCENTRIC ROD
ECCENTRIC

FIGURE 61

and the valve rod is connected to the top arm of the rocker. The rocker shaft rotates in the rocker box, which is rigid, and the motion of the eccentric is thus indirectly imparted to the valve; that is, the motion of the eccentric is reversed by the rocker arm. This is termed indirect valve gear and is the standard type in this country, although there are some engines fitted with direct valve gear in which both arms of the rocker shaft extend upwards, or in the same direction, in which case the motion is not reversed.

The valve is connected to the valve stem usually by a yoke or frame that loosely embraces the top of the valve which is formed to receive it. This allows the valve to change its position vertically, with respect to the valve stem, as the face of the valve and the seat wear away. Sometimes the valve is secured by nuts that engage with a thread on the stem. In any case it is essential that the valve should have a small amount of freedom in its connection with the stem in order to guard against its becoming cocked or tilted on its seat, thus allowing the steam to blow past it.

The cut Fig. 61 shows the link in full forward gear; that is, the full throw of the go-ahead eccentric A affects the link block and rocker arm and through these the valve. By throwing the reverse lever back into the extreme back notch of the quadrant the link will be raised until the pin that connects the lower or backing eccentric blade to the link will be in line with the pin of the link block, which will then be affected by the full influence of the back-up eccentric B, and the engine will run backward.

It has been previously stated that the lead was increased by bringing the reverse lever nearer the center notch of the quadrant, or in common phraseology "hooking her back," in order to cause cut-off to take place earlier, and that the increase in lead was due to the radius of the link. The radius of the link is the distance, on a horizontal line, from the center of the main driving shaft, which carries the eccentrics, to the center of the rocker shaft. Ordinarily an increase in the lead is obtained by moving the eccentrics ahead on the shaft, but in this case the eccentric straps are moved back on the eccentrics by raising the link, as will readily be seen by a study of the diagram, Fig. 61. The nearer the center of the link—that is, the center of the saddle pin—is brought to the center of the link block, the more are the eccentric straps moved back upon the eccentrics and the shorter will be the cut-off and the greater the lead.

With locomotives having very long eccentric blades there will not be such a marked increase in the lead as with those having very short blades, for the reason that the short rods will cause the straps to move farther around and back on the eccentrics by raising the link to which the ends of the rods are connected, whereas if the rods were longer their ends at the link could be raised considerably and still not materially affect the positions of the eccentric straps. This would indicate that in setting the valves of a locomotive, when it comes to adjustments for lead, attention should be paid to the radius of the link, which, as before stated, is the distance from the center of the main driving shaft to the center of the rocker shaft.

Authorities differ in regard to the amount of lead that locomotive valves should have at full gear. The older practice was to give an eighth of an inch, but of late years the tendency has been to cut it down to a sixteenth, or a thirty-second, and some authorities recommend even negative lead, which means no lead at full gear. They claim that too much lead is detrimental to an engine, causing more wear and tear to the valve gear, also that the preadmission of steam is too great at mid-travel. With a 4-ft. radius the valves should be set line and line, (no lead) forward and back gear. With a 6-ft. radius one-sixteenth of an inch is required, and with a radius of 8 ft. the valves should have one-eighth of an inch positive lead forward and back gear. The travel of the valve is also reduced by hooking the reverse lever back from either full gear towards the center notch of the quadrant.

Questions

152. What is a simple engine?
153. What is a compound engine?
154. Why is a compound engine more economical in fuel than a simple engine is?
155. Describe the saddle plate of a locomotive.
156. How is this casting secured to the smoke box?
157. What supports the cylinders?
158. How does the steam pass from the steam pipes in the smoke-box to the valve chests?
159. Describe the steam chest.
160. Who invented the D slide valve?
161. How many functions must a slide valve perform?
162. What is the first of these?
163. What is the second function?
164. Describe the third function.
165. What must the valve do in the fourth function?
166. What is the fifth function of the slide valve?
167. What is outside lap?
168. What is inside lap or exhaust lap?

169. What is inside clearance or exhaust lead?

170. What is the purpose of inside lap?

171. How much inside lap is usually given a locomotive slide valve?

172. What is the purpose of inside clearance?

173. What class of engines is it used on?

174. Do the valves of locomotives in ordinary service have or need inside clearance?

175. What is meant by cut-off, as applied to a slide valve?

176. What is compression?

177. When does compression begin?

178. What advantage is there in compression?

179. When does release occur?

180. What is steam lead?

181. What is meant by valve travel?

182. What is over travel?

183. What are the objects aimed at in giving a valve lead?

184. What would be the result if a valve had no lead?

185. Name another bad effect that would occur if a valve had no lead?

186. What is meant by clearance?

187. What is mechanical clearance?

188. What causes the lead on a locomotive valve to increase when the reverse lever is hooked back towards the center notch?

189. What does the term valve gear include, as applied to a locomotive?

190. What kind of a valve gear does a locomotive require?

191. Why are two eccentrics needed on each side of a locomotive?

192. What is the link saddle?

193. How is the link and the ends of the eccentric rods that are connected to it supported?

194. For what purpose is the link block?

195. What is an indirect valve gear?

196. What is a direct valve gear?

197. How is the slide valve usually connected to the valve stem?

198. What other method is sometimes used?

199. Why should a slide valve have a small amount of freedom in its connection?

200. What effect does it have upon the cut-off when the reverse lever is brought back towards the center notch?

201. What is the radius of the link?

202. Explain why it is that when the reverse lever is hooked back the lead increases.

203. How does this affect locomotives having long eccentric blades?

204. About how much lead is usually given an engine at full gear?

205. How should the valve be set when the radius of the link is 4 ft.?

206. What should the lead be with a 6-ft. radius?

207. How much lead should be given the valves when the radius is 8 ft.?

208. How is the travel of the valve affected by hooking the reverse lever back?

CHAPTER V

VALVE SETTING

As considerable time has been devoted to a study of the mechanism by which the valves of a locomotive are operated, it is now in order to take up the subject of valve setting, a subject which every young man who is ambitious to become a successful locomotive engineer should endeavor to thoroughly familiarize himself with. In fact, such knowledge is becoming more and more a necessity each year. This is indicated by the increasingly rigid examinations to which applicants for promotion from firemen to engineers are subjected.

The correct setting of the valves of a locomotive means that the adjustment of the positions of the eccentrics on the driving axle and the lengths of the eccentric blades, valve rods and valve stems is such that each valve will give the required distribution of steam to the piston that it is to serve. This has already been explained under the heading, Function of the Slide Valve. As the great majority of locomotives are equipped with indirect link valve gear, attention will now be directed to the setting of valves operated with this type of valve gear. One of the first things to be done is to see that the driving wedges are properly adjusted, also that the main rod keys at both ends are correctly tightened. It is also well to see that the eccentric rods are connected in the right way, which means the go-ahead eccentric rod to the top end of the link and the back-up eccentric rod to the bottom end of the link.

Don't forget that with the indirect link motion the eccentric that controls the valve always follows the crank pin. That is, when the pin is on the forward center, for instance, the body of the go-ahead eccentric will be above the axle and that of the backing eccentric will be below, and both eccentrics will be advanced towards the pin sufficient to overcome the lead and lap of the valve. This is termed angular advance of the eccentric. The eccentrics should be placed as near as possible in these positions and the set screws slightly tightened. Of course the positions of the eccentrics can only be guessed at on the start. Their correct positions on the driving shaft can only be arrived at after the dead centers have been located. The reverse lever should also be tested to see that the latch will enter each extreme notch.

The next most important proceeding is to get the port marks properly located on the valve rod. This, of course, must be done while the steam chest cover is off. The valve stem key should be examined to see that it is securely tightened. Next examine the back end of the valve rod and see that it will connect with the rocker arm without cramping or twisting the valve stem, which would be liable to throw the ends of the yoke up or down, thus cramping the valve. As the steam chest cover is off, the chest itself should be firmly clamped to the cylinder by screwing some of the nuts down upon washers or bushings, being careful not to mar the copper joint on top of the chest. The valve stem gland should be in place and the valve rod connected up in order to keep the stem at its proper height, as any variation in the height of the stem will cause an error in the use of the tram. If there is any lost motion between the valve and the valve yoke (and there should be a little), it should, while getting the port marks, be taken up by the use of liners between the back of the valve and the yoke, as shown at A in Fig. 62. Next move the valve back just far enough to permit a piece of thin tin to be inserted between the edge of the forward port and the forward edge of the valve at point V. The valve is now in the correct position for the forward port mark to be placed upon the valve rod; so with a prick punch make a small center at C on the cylinder, and from this point with the valve tram, as shown in the cut (Fig. 62), scribe the line F on the valve rod. Next remove the liners and place them at the front of the valve at A, Fig. 63, and

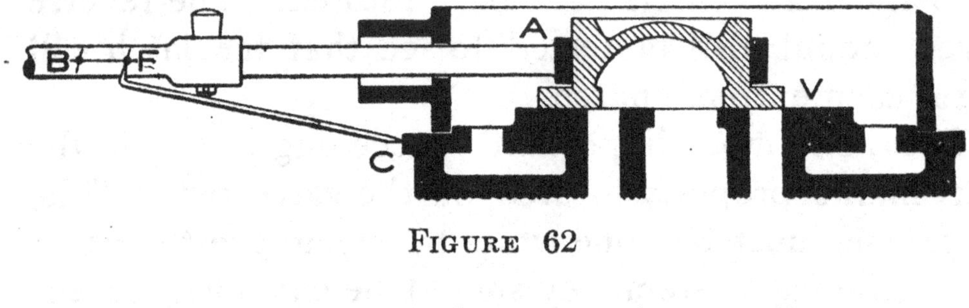

FIGURE 62

move the valve ahead far enough to allow of the insertion of the piece of tin between the edge of the back port and the back edge of the valve, as shown at V, Fig. 63. Now take the tram and from point C scribe the line B on the valve rod. Next, using a box square, scribe a parallel line on the valve rod, and at the two points where lines F and B intersect the parallel lines, make two small centers. Center F is the forward port mark and center B is the back port mark.

The mid-travel or central position of the valve on its seat should also be marked, which is done in the following manner: With a pair of dividers find the exact center between points F and B and at this point make another small center M, Fig. 63. This point will represent the central position of the valve. The points F and B indicate the points of admission and cut-off, and the distance from F to M or from B to M equals the lap of the valve. If the valve has neither inside lap nor inside clearance the point M will represent the points of both release and compression. If the valve has inside lap or inside clearance it will be necessary to locate two additional points on the valve rod. These points may be found in the following manner: Set a small pair of dividers to a distance equal to the inside lap or inside clearance, whichever the valve has, and from the center M describe a small circle, and the two points where the parallel line on the valve rod bisects this circle will indicate the points of release if

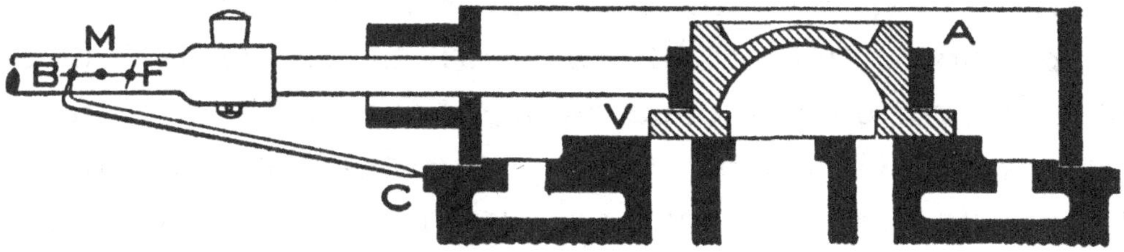

FIGURE 63

it is inside clearance or of compression if it is inside lap. These two points are very seldom used in practice, but if, owing to the construction of the valve, it should become necessary to use them it should always be remembered what they represent, whether inside lap or inside clearance.

The next important move in valve setting is to find the four dead centers. It is very important that the dead centers be accurately located. Although the crosshead moves very little while the crankpin is near the dead center, yet the valve is moving at nearly its greatest speed, being at about half travel, and a very slight error in locating the dead centers will seriously affect the accuracy of the whole work.

The term dead center is commonly taken to mean that the driving wheels are in such a position that the centers of the driving axles and the centers of the crankpins are in a horizontal line, but this is not always the dead center. Theoretically the term implies that the center of the crosshead pin, the center of the crank pin, and the center of the driving axle be exactly in line, regardless of whether that line be horizontal or inclined. Therefore the crankpin must pass two dead centers in each revolution, viz., the forward dead center and the back dead center. Consequently there are in locomotive valve setting four dead centers to be located and marked: first, the right forward dead center; second, the right back center; third, the left forward center, and fourth, the left back center. It makes no particular difference which center is found first, but for convenience the right forward center may be taken. Of course finding the dead centers of a locomotive implies that the driving wheels are to be revolved more or less. This means that the engine must be pinched ahead or back as required, which involves considerable labor on the part of helpers, as many an engineer who has served an apprenticeship in the shop can testify. Many well conducted shops are equipped with roller devices of various designs which are placed under the drivers to be revolved. Such a machine is illustrated in Fig. 64. It may be operated by one man by means of the lever shown. In some up-to-date shops the rollers for moving the drivers are operated by small air engines.

And now to find the right forward dead center. Turn the driving wheels forward until the crosshead is within an inch of the forward end of the stroke, as indicated in Fig. 65. Then, having first examined the wheel cover to see that it is securely fastened, make a center at any convenient point on it, as at C; also make a center at point F on the forward guide block. Now, using a short tram called a cross head tram, describe from point F an arc G on the cross head; also

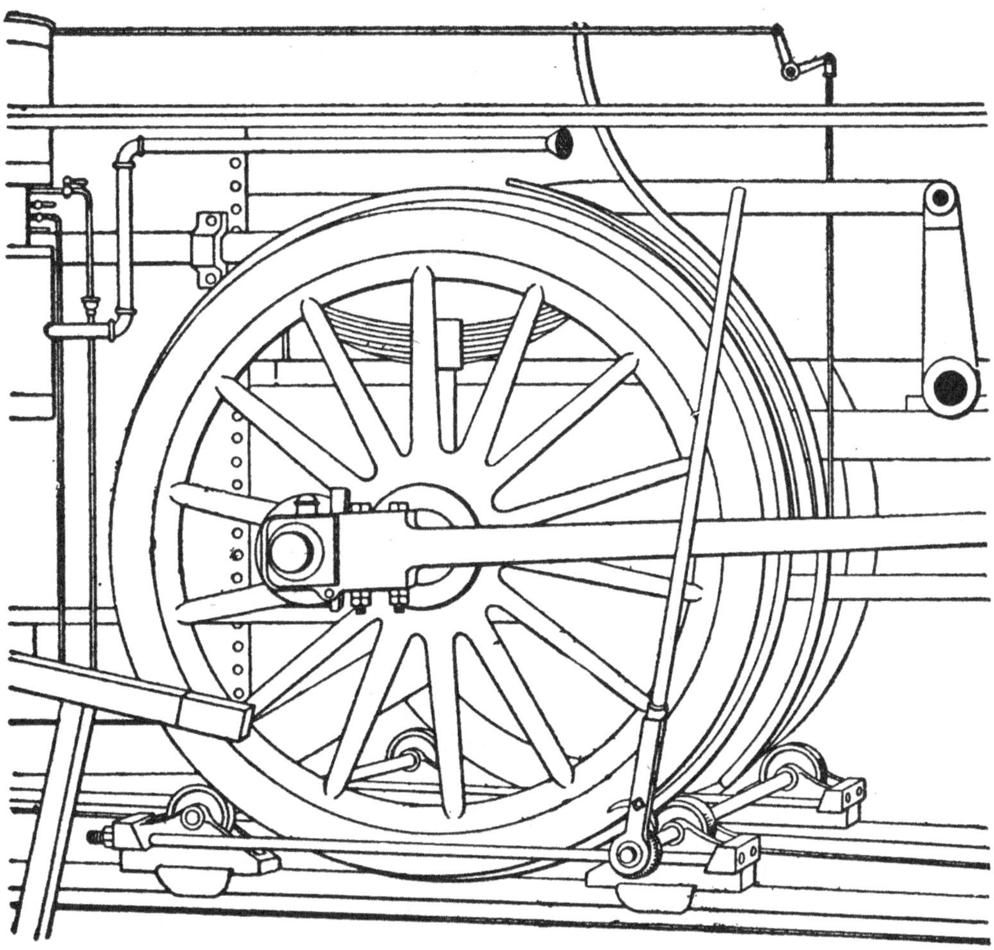

FIGURE 64

with one point of a longer tram, called a wheel tram, set in center C describe the arc A on the tire of the wheel. Next turn the wheel ahead as indicated by the arrow until the cross head has passed the limit of its forward travel and has receded on its return stroke far enough to bring the arc G a short distance back of

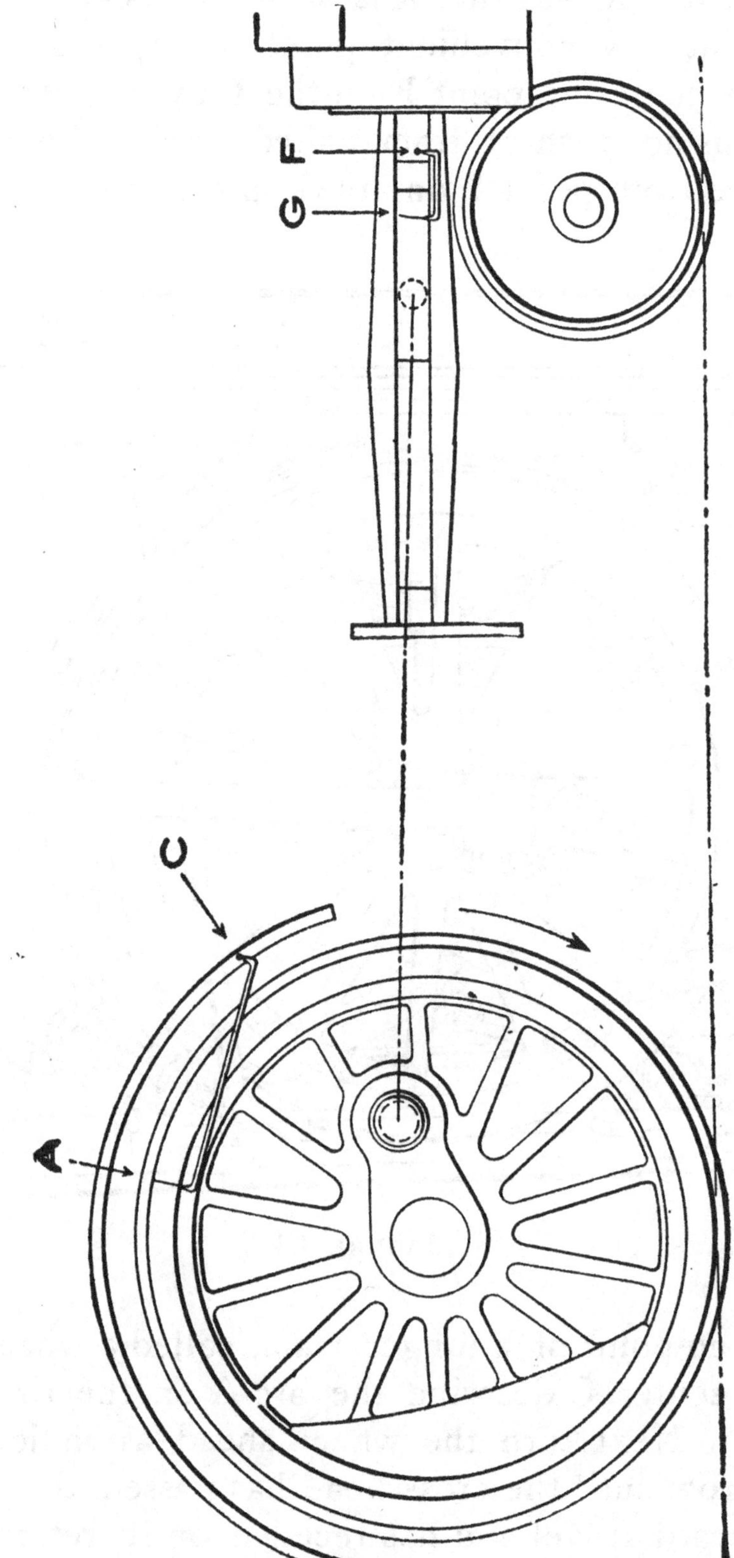

FIGURE 65

the point of the tram, one point of which is set in center F. Now reverse the motion and pinch the wheels slowly backward until the arc G comes directly under the point of the tram. Then stop, and with the wheel tram set in center C scribe an arc B on the tire, as indicated in Fig. 66. Now with a pair of hermaphrodites describe the arc D E on the tire, and at the points where the lines A and B intersect arc D E make small centers, and with a pair of dividers find the exact center between these two points. This center is indicated in the cut (Fig. 66) by the letter H. This point is the dead center, and a small circle should be drawn around it to distinguish it from the other centers.

Perhaps the query might arise in the mind of the student, why is it that in turning the wheels ahead until the pin had passed the center they were turned far enough to bring the cross head back of the position it was in when arc G was scribed? The answer is, that when arc G was scribed the pin was pushing the cross head forward and all the lost motion between pins and brasses was taken up in that direction. If, after the pin had passed the center and the crosshead was traveling back, it had been stopped at arc G, the lost motion would have been taken up in the other direction, for the reason that the pin was now pulling instead of pushing the crosshead. The result of this would have been an error in the location of the arc B and also of the point H. But by pulling the crosshead back past arc G and then reversing the motion and allowing the pin to push the crosshead until the dead point H was located, the lost motion was taken up in the same direction as when arc G was first drawn.

Having now found the dead center at point H, the next move is to throw the reverse lever into the

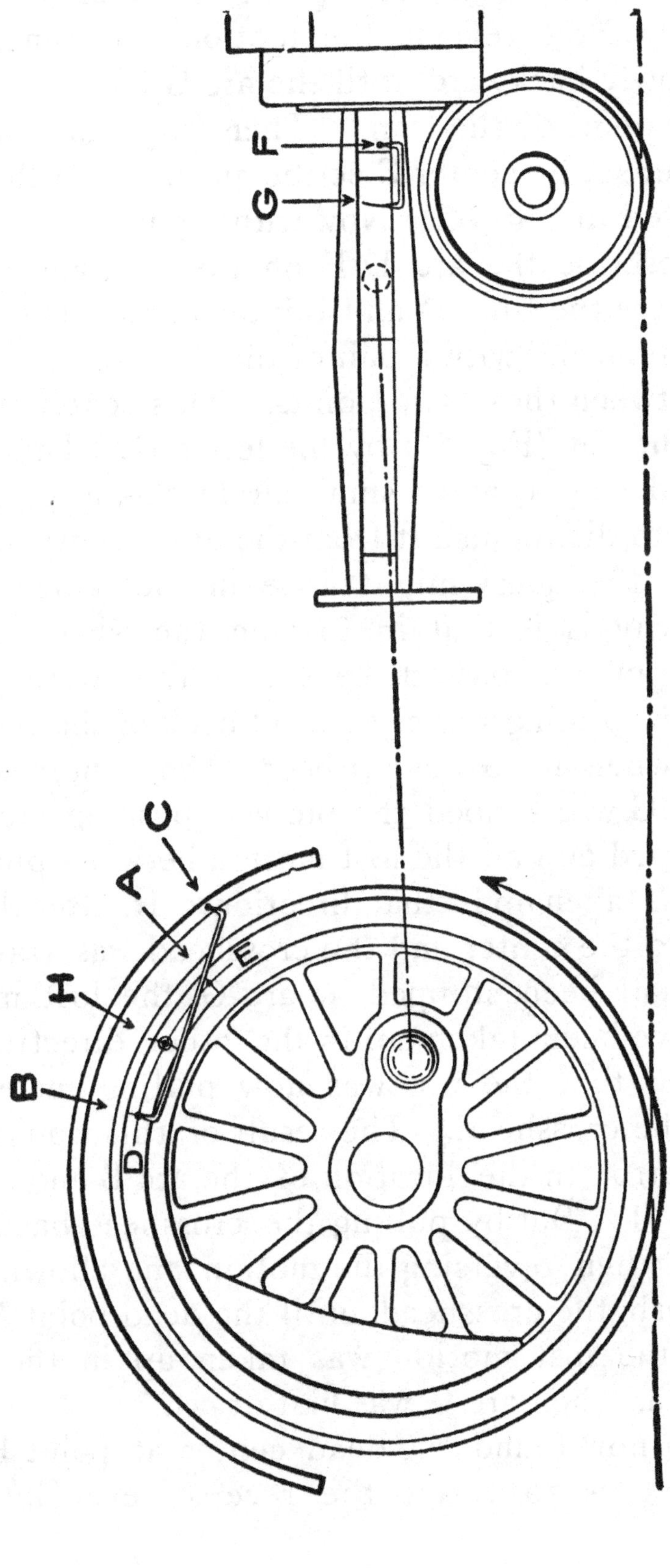

FIGURE 66

extreme back notch, so as to take up all the lost motion in the valve gear while backing up. Now start to pinch the engine back, and with one point of the wheel tram in center C, watch the center H and when it comes exactly under the other point of the tram, stop. The engine is now on the right forward dead center, and a vertical line should be scribed on the guides exactly in line with the front end of the crosshead. This line indicates the extreme forward travel of the crosshead, and it is important that it should be placed there.

While the engine is in the position it now is, that is, on the right forward dead center, and the valve gear in the backward motion with all the lost motion taken up in that direction, take the valve tram and from point C, Fig. 62, scribe an arc on the valve rod, starting slightly above the parallel line and extending considerably below it. The distance of this arc, measured on the parallel line from center F, indicates the position of the valve, as regards lap or lead for backward motion. The reason this arc is drawn below the line is that the back-up eccentric is moving the valve, and by having the arc below the parallel line it is easily distinguished from the other arc soon to be scribed for the forward motion. Now pinch the wheels back until the crankpin is about 6 in. above the dead center. Then put the reverse lever in full forward motion and pinch ahead until the pin is again on the forward dead center, and with the valve tram again set in point C scribe another arc on the valve rod, this time extending above the parallel line. The distance this arc is ahead or back of the point F indicates the amount of lap or lead the valve has in the forward motion, when the crankpin is at right forward dead center.

Before making any adjustments, go round to the left side of the engine and find the left forward dead center; also mark the left valve rod for both forward and back motion in the same manner as the right valve rod was marked. Having completed the location of the forward dead centers for both sides, the next move is to start on the right-hand side again, and pinch the engine towards the right back dead center, which is to

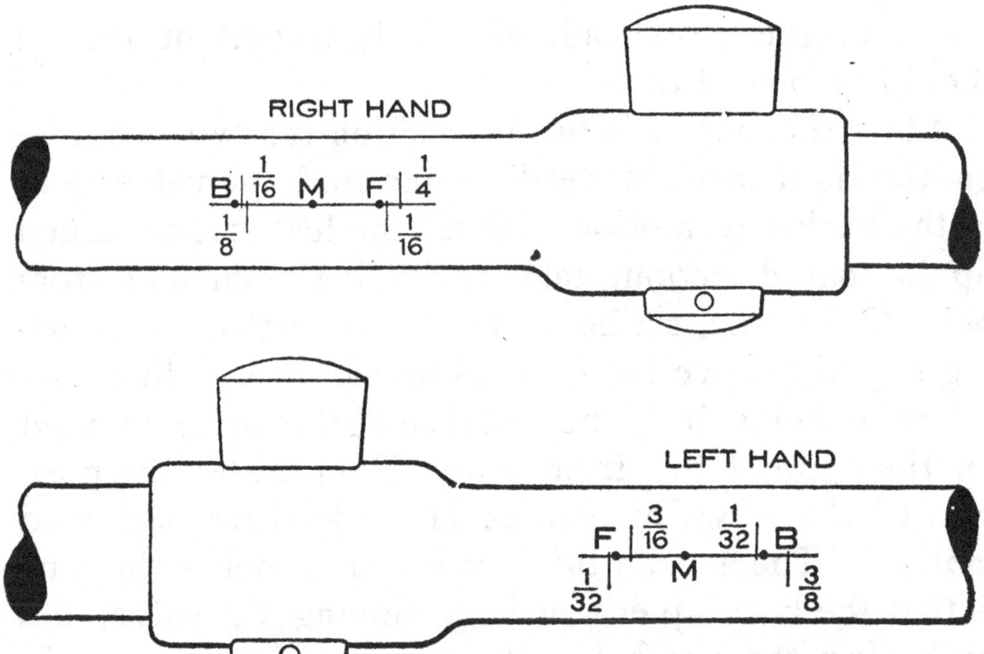

FIGURE 67

be found in the same manner as the forward one was. Next find the left back dead center, marking the valve stems and crossheads in both instances exactly as before.

The valve rods will now show a marking similar to that illustrated by Fig. 67, with the exception that the figures may not coincide, as the figures shown in Fig. 67 are merely assumed for purposes of explanation. As before stated, the arcs which have been scribed across the parallel lines indicate by their position relative to the port marks F and B whether the valve has lap or lead at either dead center. If an arc comes between the port marks it indicates lap, if outside it indicates lead. Referring to Fig. 67, the two forward motion arcs on the right side valve rod, which are distinguished from the back motion marks by being above the parallel line, show that the valve has ¼-in. lead at the forward port mark and $\frac{1}{16}$-in. lap at the back port mark.

When the valve tram reaches from center C to either center B or center F, it indicates that the valve is at the point of cut-off, and since the valve is to travel equal distances each way from these points, it follows that by measuring the distance from B or F to the arcs, it may be determined how much and whether to lengthen or shorten the eccentric blades. First take the right forward motion. The distance from F to the mark above the line is ¼ in., and from B to the mark for back motion is $\frac{1}{16}$ in., therefore the length of the right forward motion eccentric blade must be changed so as to equalize these distances, and the point to be determined is, shall it be lengthened o- shortened? This can be done in the following manner: Take a small pair of dividers and find the exact center between the two tram marks above the parallel line. If this center is ahead of center C the eccentric blade must be shortened, if back of it the blade must be lengthened. If the engine has a direct valve motion, this rule is to be reversed and the adjustments made accordingly.

The next point to be determined is, how much shall the blade be lengthened or shortened? A good rule to follow in this instance is this: When the arcs on the valve rod are both back or both ahead of the port marks F and B, the length of the eccentric blade should be altered an amount equal to one-half the sum of the distances between the port marks and the arcs, or if one arc is back and the other is ahead of their respective port marks, the length of the blade should be changed an amount equal to one half the difference of the distances between the port marks and the arcs. In this particular case the valve has traveled too far back, as shown by the $\frac{1}{4}$-in. lead on the forward port mark and the $\frac{1}{16}$-in. lap on the back port mark. Therefore the blade must be shortened one-half the sum of these distances, or $\frac{\frac{1}{4}+\frac{1}{16}}{2}=\frac{5}{32}$ in. This will square the valve for right forward motion, or in other words equalize its travel in either direction from mid position, as may be proved by the following simple calculation. The valve had $\frac{1}{4}$-in. lead at the forward port mark, the eccentric blade is shortened $\frac{5}{32}$ in., thus bringing the point of the tram that much nearer to F. Then $\frac{1}{4}-\frac{5}{32}=\frac{3}{32}$ in., which is now the lead at the forward end. At the back end, instead of lead, the valve had $\frac{1}{16}$ in. lap.

After the blade is shortened $\frac{5}{32}$ in. it will be found that the valve has been moved that distance ahead from its former position. Then by deducting the $\frac{1}{16}$ in. lap from $\frac{5}{32}$ in. change it will be found that the valve has $\frac{3}{32}$ in. lead at the back end also. It may be assumed that the valves are to have $\frac{1}{32}$ in. lead when in full gear, and as the valve under consideration now has $\frac{3}{32}$ in. at both ends, it will be necessary to reduce it by turning the eccentric back upon the shaft. However, no changes should be made until all the tram marks on both sides of the engine have been examined and a memorandum made of the changes required, as, for instance, R. F. Ecc., shorten blade $\frac{5}{32}$ in., $\frac{1}{16}$ in. lead off.

The tram marks for the right backward motion should be examined next. These marks are below the parallel line, and measurements show that the valve has $\frac{1}{16}$ in. lead at the forward end and $\frac{1}{8}$ in. lap at the back; therefore the blade of the right back up eccentric must be shortened thus, $\frac{\frac{1}{8}+\frac{1}{16}}{2}=\frac{3}{32}$ in. This will square the valve for right backward motion, but it will still have $\frac{1}{32}$-in. lap at both ends, when $\frac{1}{32}$-in. lead is required; therefore the eccentric must be turned ahead. These changes should be noted down as follows: R. B. Ecc., shorten blade $\frac{3}{32}$ in., $\frac{1}{16}$-in. lead on.

If the upper and lower rocker arms are of the same length the figures for changing the length of the eccentric blades will be all right, but if, as is often the case, the lower arm is shorter than the upper one, the length of the blades will not need to be changed quite as much as is indicated by the marks on the valve rod. But it will be assumed in this instance that the arms are of equal length, and the lengths of the eccentric blades for the right-hand side may be adjusted according to the above figures.

Next go to the left-hand side. By reference to Fig. 67 it will be seen that the valve has $\frac{3}{16}$-in. lap on the left forward motion in front and $\frac{1}{32}$-in. lap behind. In this instance the valve has not traveled far enough back, therefore the blade must be lengthened one-half the difference between these distances or $\frac{\frac{3}{16}-\frac{1}{32}}{2}=\frac{5}{64}$ in. This will equalize the lap at both ends, making it now $\frac{7}{64}$ in., and in order to obtain the $\frac{1}{32}$-in. lead desired it will be necessary to move the eccentric ahead on the shaft an amount sufficient to overcome the $\frac{7}{64}$-in. lap plus $\frac{1}{32}$-in. lead, a total of $\frac{9}{64}$ in. This is to be noted down as follows: L. F. Ecc., lengthen blade $\frac{5}{64}$ in., $\frac{9}{64}$-in. lead on.

Examination of the two left back motion marks shows that the valve has 3/8-in. lead at the back and

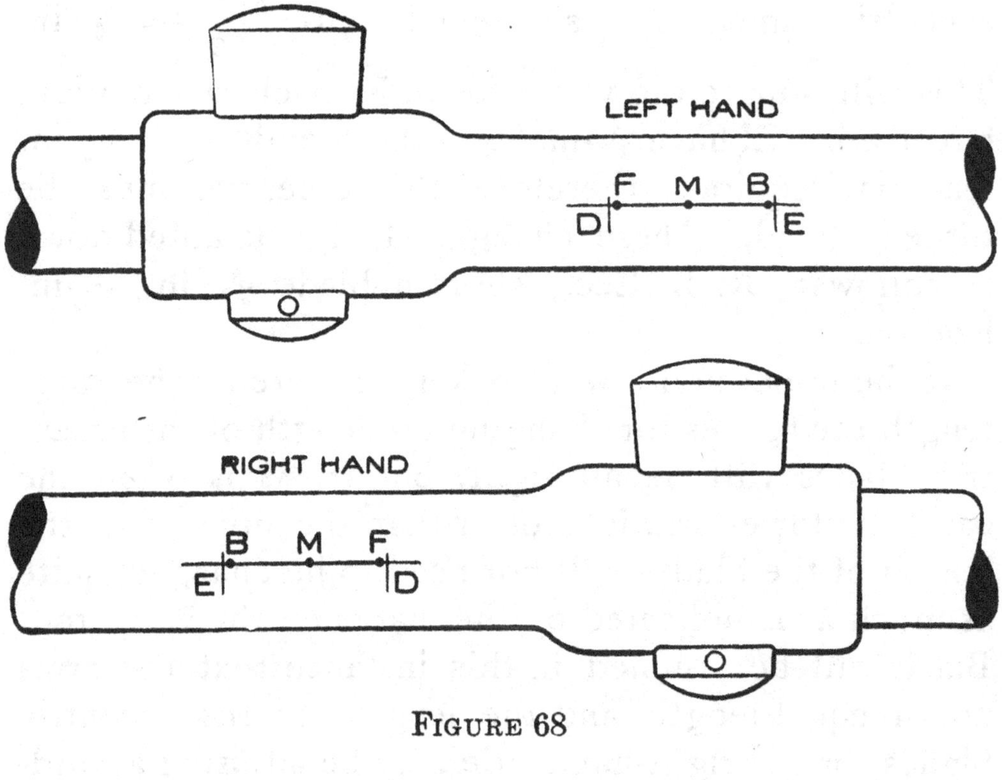

FIGURE 68

$\frac{1}{32}$-in. lead in front. Therefore the back up eccentric blade should be lengthened $\frac{\frac{3}{8}+\frac{1}{32}}{2}=\frac{13}{64}$ in. This will give the valve $\frac{13}{64}$-in. lead at both ends, but as $\frac{1}{32}$-in. lead is all that is required, it will be necessary to turn the eccentric back on the shaft far enough to overcome $\frac{11}{64}$ in. of this surplus lead. Note this down also as follows: L. B. Ecc., lengthen blade $\frac{13}{64}$ in., $\frac{11}{64}$ in. lead off.

The lengths of all the eccentric blades should now be adjusted according to the figures obtained, after which it will be in order to set the eccentrics. It is generally best to set the forward motion eccentric first, because it is easier to get at than the back motion one is; then if the backward motion eccentric needs to be changed enough to affect the lead in forward motion, the forward motion eccentric can easily be reset, and it will need to be moved so little that the backward motion will not be affected enough to require any further attention.

As before stated, it is desired to give the valve $\frac{1}{32}$-in. lead in full gear in both forward and backward motion, and before setting the eccentrics it will be necessary to have lead marks on the valve rods for a guide. To get these marks, set a pair of dividers to the distance between the centers B and F, Fig. 68, plus the lead, in this instance $1\frac{1}{2}$ in. + $\frac{1}{32}$ = $1\frac{17}{32}$ in. Then with one point of the dividers in center F scribe an arc E across the parallel line, back of center B, also from center B scribe an arc D in front of F. These points, E and D, will serve as guides in setting the eccentrics for lead. The next move is to place the engine on the dead center; either one will do, but for convenience it may be assumed that it is the right forward dead center. When adjusting the lengths of the eccentric blades it was found that with the engine and reverse lever in this position the valve had $\frac{3}{32}$-in. lead. This must be reduced to $\frac{1}{32}$ in., and it might be done by simply turning the right forward motion eccentric back upon the driving shaft, but that would take up the lost motion in the opposite direction to what it is when the engine is running ahead, and this would cause an error in the working of the valve. The proper method is to turn the eccentric backward far enough to take off all the lead, and then turn it slowly ahead until the valve tram will reach from center C, Fig. 62, to the lead mark D, Fig. 68. Next throw the lever into full back gear and proceed to set the right backward motion eccentric.

After getting the right backward motion eccentric blade adjusted to the correct length it was found that the valve had $\frac{1}{32}$-in. lap at both ends; therefore this eccentric should be slowly turned ahead on the shaft until the tram will reach from center C to lead mark E, Fig. 68. This will square the right-hand valve and give it the desired lead, and the next move is to go around to the left side, throw the reverse lever into full gear ahead again and pinch the engine onto left forward center.

After adjusting the left forward eccentric blade to the correct length it was found that the valve had $\frac{7}{64}$-in. lap. The eccentric must therefore be turned ahead until the tram will reach from center C on the left cylinder to lead mark D on the left valve rod. Next proceed to set the left backward motion eccentric, and in doing this the wheels should be pinched ahead about 6 in., then place the reverse lever in full back gear and pinch the engine back onto the center. This is done to take up all the lost motion in the direction in which the engine is to run—a very important matter that should never be lost sight of in working from the dead center for either forward or backward motion. After getting the left backward motion eccentric blade the right length the valve had $\frac{13}{64}$-in. lead at both ends of the stroke. The eccentric should therefore be turned back sufficiently far to take off all the lead. Then with all the lost motion taken up, turn the eccentric slowly ahead until the tram point will drop into lead mark E, Fig. 68.

The engine is now square, and the valves have the correct amount of lead all around. The eccentrics should be securely fastened in their proper location either by set screws or keys, and it will be next in order to ascertain the points of cut-off, so that they may be equalized as near as possible, for be it remembered that no matter how accurate the valves may have been set, as regards lead, travel, etc., they very seldom cut off the steam at the same distance from the commencement of the stroke at each end of the cylinder, and one cylinder may be getting more steam than the other. This is due to the fact that the link motion is not a perfect valve gear, various errors being introduced by the angularity of the main rod, and eccentric rods, and the off-set of the link pin holes from the link arc, but these errors can be almost entirely eliminated by making certain changes, among which may be mentioned the off-setting of the link saddle stud, although with case-hardened links and the saddle rigidly bolted to the link this method is not always practicable.

Another very common method is to equalize the forward motion by changing the length of the backward motion eccentric blades, thus sacrificing equality of lead and cut-off in the back gear, but as a locomotive does the greater portion of its work in the forward gear, except it be a switch engine, this plan is permissible.

Another method employed to some extent is to sacrifice equality of lead in both forward and back gear for equality of cut-off. But before either plan can be adopted it will first be necessary to find the points of cut-off as the valves are now set. As a locomotive engine performs the principal part of its work with the reverse lever hooked back towards the center notch and the valves cutting off at early points in the stroke, it is more important that the steam should be equally distributed with the lever in the working notch than with it down in the corner.

Passenger engines usually cut-off at from 4 to 6 in., and freight engines at from 6 to 9 in. As in setting the eccentrics, a start at finding the points of cut-off may be made with the engine on either dead center, but for convenience it may be assumed in this instance that the engine has been placed on the right forward dead center. First try the cut-off in backward motion. Pinch the wheels backward until the crosshead has traveled about 6 in. from the extreme travel mark on the front end of the guides. Then stop the motion, and with the point of the valve tram in center C, Fig. 62, move the reverse lever back of the center until the tram will drop into the forward port mark F. Put the lever one notch farther back, then pinch the wheels backward until the tram again drops into the forward port mark F, thus indicating that the point of cut-off has been reached. Now measure the distance from the front travel mark to the front end of the cross head. Suppose it is found to be 7½ in. Chalk this down on the front end of the outside guide. Use the outside guide for the backward motion, because the backward motion eccentric is on the outside. Now pinch the wheels farther back until the steam is cut off on left side back end of cylinder, which can be ascertained by the use of the tram in the same manner as on the right side. It may be assumed that cut-off takes place when the piston has traveled 8¾ in. from beginning of stroke. Now turn the wheels still farther back, until the right pin passes the front center and reaches the point where cut-off takes place, which will be assumed to be 8 in. from commencement of stroke. These figures should all be marked down with chalk on the outside guides for the backward motion as they are found, and the reverse lever must be left in the same notch until all four points of cut-off for backward motion are located. Next pinch the engine still farther back until the left pin passes the front center and cut-off for this end is reached, which may be taken at 9 in. for the present.

These investigations show that cut-off for the right cylinder occurs at 7½ in. of the backward and 8 in. of the forward stroke, and that for the left cylinder cut-off takes place at 9 in. of the backward and 8¾ in. of the forward stroke. These figures indicate that the right-hand valve is traveling a little too far ahead and the left valve a short distance too far back, and the cut-off for each side may be equalized by slightly changing the lengths of the backward motion eccentric blades; that of the right-hand one must be lengthened and the left one will need to be shortened, and how much to change them may be found as follows:

Taking the left side first, cut-off occurs on the front end of cylinder at 9 in. and on the back end at 8¾ in., and the average is $\frac{8\frac{3}{4}+9}{2} = 8\frac{7}{8}$ in., which is the distance from each end of the stroke at which cut-off will occur when it is equalized. Now pinch the wheels forward enough to bring the crosshead 8⅞ in. from the end of the stroke and enough more to take up all the lost motion when turning back. Next pinch the engine backward until the crosshead is again 8⅞ in.

from the beginning of the stroke at the front end, and with the valve tram in center C, Fig. 62, scribe a mark on the valve rod. This mark will be a short distance ahead of center F, and this distance shows how much too far back the valve is traveling, and the eccentric blade must be shortened enough to throw the valve ahead that much. This will equalize the cut-off for the left side in the backward motion, and the right side should be treated in the same manner, except that in this case the backward motion eccentric blade must be lengthened, because the valve was traveling too far ahead, the cut-off for the forward stroke being 8 in. and for the backward stroke 7½ in., and the average is $\frac{8+7\frac{1}{2}}{2} = 7\frac{3}{4}$ in., which will be the point of cut-off for the right side in backward motion when the proper change is made.

This will leave considerable difference between the two sides, the cut-off on the left side occurring at 8⅞ in. and on the right side at 7¾ in., but this will be remedied later on, and the next move will be to equalize the cut-off for the forward motion by commencing with the backward stroke on the right-hand side. Pinch the engine ahead until the pin passes the forward center and draws the crosshead back 6½ in. from the beginning of the stroke. Move the reverse lever ahead to the corner, then move it slowly backward until the valve closes the port, as will be indicated by the valve tram when it reaches from center C to center F, Fig. 62; then put the lever in the first notch ahead of that position and leave it there until the points of cut-off have all been found for the four strokes. Now pinch the engine ahead until the tram again shows that the point of cut-off is reached. This may be assumed to be 8 in. back of beginning of the stroke. Mark this down on the front end of right inside guide, then turn the wheels ahead and get the cut-off for the front end of left-hand cylinder, which will be, say 7 in. Again pinching ahead, find the cut-off for the back end of the right-hand side to be 8¾ in., and still turning ahead, find cut-off for back end of left side to take place at 8 in.

For convenience, the cut-off for the left side may be equalized first. It was found that cut-off occurred at 7 in. for the backward stroke and at 8 in. for the forward stroke, the average being $\frac{7+8}{2} = 7\frac{1}{2}$ in., and in order to equalize the travel of the valve, which now travels too far ahead, it will be necessary to lengthen the eccentric blade. Pinch the wheels back far enough to bring the crosshead within less than 7½ in. of the end of the stroke, so that when turned ahead again all lost motion may be overcome. Now pinch ahead again until the crosshead is exactly 7½ in. from the beginning of the stroke, and with the valve stem tram in center C scribe a mark on the valve rod, and the distance of this mark from center B is the amount that the eccentric blade must be lengthened. This will equalize the cut-off for forward motion on the left side, and the right side next demands attention. Here the point of cut-off for the backward stroke was 8 in., and for forward stroke 8¾ in., the average being $\frac{8+8\frac{3}{4}}{2} = 8\frac{3}{8}$ in., which is the point at which cut-off for forward motion on the right side must be equalized for the present.

The right forward motion eccentric blade will also need to be lengthened, as the valve travels too far ahead, and the correct amount to lengthen the blade may be found in the same manner as with the left side. This will leave the points of cut-off for forward motion as follows: for right-hand side, 8⅜ in.; left side, 7½ in. In backward motion, as equalized, cut-off for right side is 7¾ in., left side 8⅞ in. It will thus be seen that in forward gear cut-off is earliest on left side and in back gear it occurs latest on that side. In order to overcome this unequal condition one of two things may be done, either lengthen the link hanger on the left side or shorten hanger on the right side. The former method will be adopted, but before making any alterations it will be necessary to ascertain the amount to lengthen, and this may be done in the following manner:

Put the reverse lever in the same notch that it was in when the cut-off in forward gear was found, and measure the distance from any stationary point directly above or below the upper link hanger pin on the left side to the center of that pin. Now pinch the engine ahead far enough to bring the left crosshead the same distance from the beginning of the stroke as the right crosshead was when cut-off took place, which distance is 8⅜ in. This is where cut-off must occur on the left side also. Now move the reverse lever ahead about four notches, and then with the point of the valve tram in center C move the lever slowly back until cut-off occurs as indicated by the tram. Now measure the distance again, from the same stationary point to the center of the upper hanger pin. The difference between this distance and the distance between these two points as first found is the amount the left hanger must be lengthened to equalize the cut-off on both sides, or raising the tumbling shaft box slightly more than this on the right-hand side would bring about the same result as shortening the link hanger on that side. This change will slightly affect the operation of the valves in back gear, for this reason: The nearer the link block is to the center of the link, the shorter will be the cut-off, and the change made, viz., lengthening the hanger, while it throws the block farther below the center of the link in forward gear, thus delaying the cut-off, at the same time brings the link block nearer the center of the link in back gear, thus accelerating the cut-off, and this is the result wished for to cause the two sides to cut off nearer equal in back gear, as it will be remembered that cut-off on the left side occurred at 8⅞ in. in the back gear and 7¾ in. on the right side in back gear. The amount that the hanger has been lengthened may not exactly equalize the cut-off in back gear, but it will bring it near enough for all practical purposes, for the reason that the engine does very little work in back gear. Owing to the space occupied by the piston rod in the back end of the cylinder, the cut-off should occur ¼ or ⅜ in. later in the back end than in the front end of the cylinder if it is desired that the same volume of steam be admitted to each end of the cylinder.

The next points to be determined relate exclusively to the exhaust opening and closure with reference to release and compression. These events, as has been already explained, are controlled by inside clearance and inside lap. If a valve is line and line inside, having neither inside clearance nor inside lap, the point M, Fig. 63, will indicate both the opening and closure of the exhaust, but if a valve has inside clearance, release will occur before the valve has reached its central position; or if the valve has inside lap, closure of the exhaust passage will occur before the valve has reached its central position.

Now in order to ascertain at what point in the stroke either one of the above named events takes place, use a pair of small dividers and from center M describe on each valve rod a small circle, the radius of which equals the inside lap or inside clearance, as the case may be, and make two small centers where the circle crosses the horizontal line, also mark each with some distinguishing mark to show whether it represents inside lap or inside clearance.

Having gotten these marks properly located, proceed to test each event by the same method as with the cut-off, marking down the point in each stroke at which the event, be it release or compression, begins, after which compare the figures, and the changes required may be made in the same way as with the cut-off. Equalizing the cut-off incidentally affects exhaust closure, and as compression is of more importance than release, it should be made as near perfect as possible. There is, however, but one method by which the various events in the working of a valve may be made thoroughly clear, and that is by the use of the indicator.

The maximum port opening and maximum travel of the valve may be found thus: Place the reverse lever in full gear; that is, "down in the corner." Then pinch the wheels one complete revolution, and with the valve tram in center C, mark the extreme travel of the valve in each direction. The distance between the extreme points indicates the maximum travel of the valve, and the distance from either extreme point to the port mark indicates maximum port opening.

The minimum travel and minimum port opening

may be found by placing the reverse lever in the center notch of the quadrant, and then repeating the operation of turning the wheels one revolution, while at the same time the distances are noted with the tram in the same manner as before.

Questions

209. What does the correct setting of the valves of a locomotive mean?

210. What two very important details should be looked after first when preparing to set valves?

211. What should be done regarding the eccentric rods?

212. With indirect valve gear, what is the position of the eccentric that controls the valve?

213. What is meant by angular advance of the eccentrics?

214. What should be done with the reverse lever before commencing to set valves?

215. What is the next most important proceeding in valve setting?

216. How should the valve rod connect with the rocker arm?

217. If the valve rod should be cramped or twisted, how would this affect the valve?

218. What should be done with the steam chest and valve stem gland?

219. What should be done with the lost motion between the valve and valve yoke while getting the port marks?

220. Where should the port marks be placed for convenience?

221. Where will the point indicating mid travel or central position be?

222. How is the lap indicated by the marks that are now on the valve stem?

223. If the valve has neither inside lap nor inside clearance, what point indicates release and compression?

224. If the valve has inside lap or inside clearance, how may they be measured and properly marked on the valve stem?

225. What is the next important move in valve setting after getting the port marks?

226. What is the meaning of the term dead center as applied to an engine?

227. How many dead centers must the crankpin pass in each revolution?

228. How many dead centers are to be located and marked in setting the valves of a locomotive?

229. Describe in general terms the method of locating and marking a dead center.

230. How should the guides be marked while the engine is on dead center?

231. How are the marks for lap or lead located on the valve rod?

232. What should be done before making any adjustments?

233. When marks for lap or lead are located on the valve stem, how are they distinguished from each other?

234. How are the eccentric rods adjusted as to length?

235. Give the rule for finding out how much the eccentric blade must be lengthened or shortened in order to get the correct travel for the valve.

236. Suppose the valve has $\frac{1}{4}$-in. lead on the forward port and $\frac{1}{16}$-in. lap on the back port, how much must the blade be shortened?

VALVE SETTING

237. If a valve has $\frac{1}{16}$-in. lead at the forward port and ⅛-in. lap on the back port, what should be done with the eccentric rod?

238. After the lengths of all the eccentric blades have been adjusted so as to give the valve correct travel, what comes next?

239. Which eccentric should be set first?

240. Why?

241. What precaution should be taken when turning the eccentrics ahead to increase the lead?

242. After getting the eccentrics in their correct position and firmly secured, what is the next move in valve setting?

243. Mention some of the causes for variation in the cut-off.

244. How may this variation in the cut-off be equalized?

245. Mention another very common method of equalizing the cut-off.

246. Where should the reverse lever be placed while equalizing the cut-off?

247. At what point in the stroke do passenger engines usually cut off?

248. At what point in the stroke do freight engines usually cut off?

249. By what means may the point of cut-off be ascertained?

250. Suppose it is found that for the left-hand cylinder cut-off occurs at 9 in. of the backward and 8¾ in. of the forward stroke, at what distance from each end should it occur when equalized?

251. What must be done to bring about this equalization?

252. If, after equalizing the cut-off on both sides, it

is found that on the right-hand side it occurs at 8⅜ in. in forward gear and 7¾ in. in back gear, and that on the left side cut-off occurs at 7½ in. in forward and 8⅞ in. in back gear, what may be done to overcome the difficulty?

253. How may the amount to lengthen or shorten the link hanger be ascertained?

254. Why should cut-off occur a little later in the back end than in the front end of the cylinder?

255. What are the next points to be determined?

256. If a valve has inside lead or clearance, how will release be affected?

257. How may the points of release and compression be ascertained?

258. How does equalizing the cut-off affect exhaust closure?

259. Which is the more important, release or compression?

260. How may the maximum port opening and maximum travel of the valve be found?

261. How may the minimum port opening and minimum valve travel be found?

CHAPTER VI

PISTON VALVES AND BALANCED VALVES

Hitherto the plain D slide valve alone has been considered in the discussion of the subject of valves and valve setting.

There are, however, many other types of valves in use on locomotives, including piston valves, balanced slide valves, ported valves, roller balanced valves, etc.

Some of these possess many merits of their own, while others have very few points to recommend them.

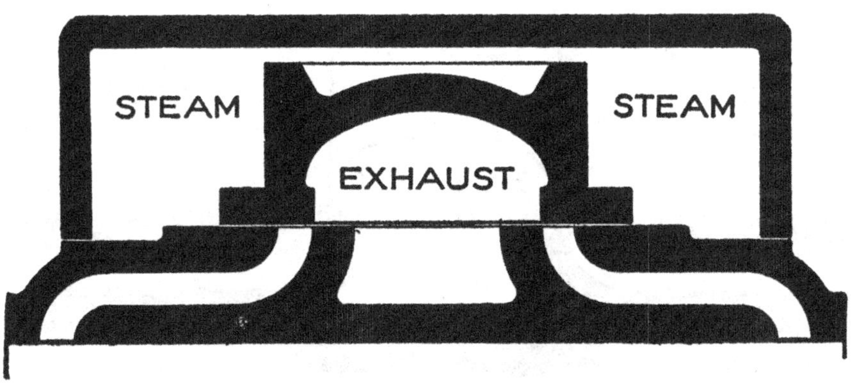

FIGURE 69

The principal objection to the use of the D slide valve is the large amount of friction caused by the action of the steam pressing the valve against its seat, and inventors have racked their brains for many years in efforts to produce a valve that would work without friction, and at the same time give a correct distribution of the steam to and from the cylinders.

The piston valve, while practically balanced, owing to the pressure of the steam acting upon each end, is, nevertheless, not a perfectly balanced valve unless the valve rod extends through both ends of the valve chamber, and this necessitates an extra gland and set of rod packing. In order to more clearly illustrate this idea, reference is made to Figs. 69 and 70. Fig. 69 shows a plain D slide valve, and it will be noticed that the full pressure of steam in the valve chest acts upon the back of the valve. Of course there is a certain amount of back pressue from the steam port and exhaust port that tends to overcome the direct pressure; still there is an enormous strain on the valve gear that is required to move a valve under such conditions.

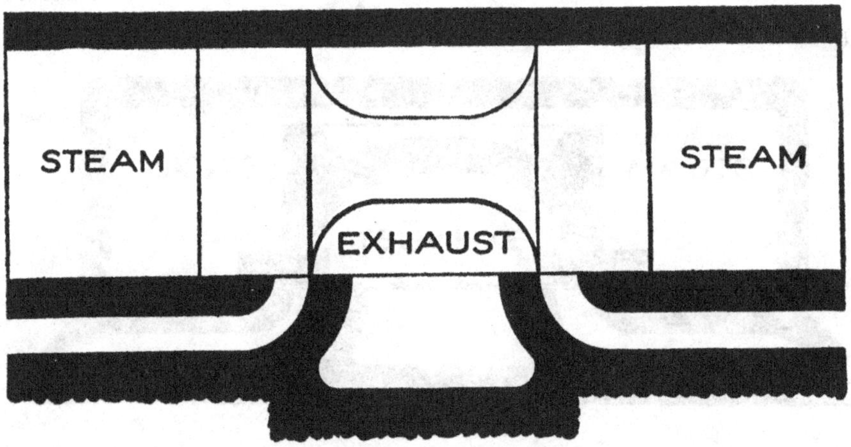

FIGURE 70

Fig. 70 shows a solid piston valve with outside admission, being thus identical in action with the D valve.

No valve rod is shown in either cut, but it will easily be seen that with the valve rod attached to but one end of the piston valve the area of that end will be decreased just so much, and the valve will be unbalanced by an amount equal to the sectional area of the valve rod, but this amount is so insignificant that builders very seldom add the extended valve rod, and so the piston valve may be considered as balanced, the only friction being that due to the weight of the valve and the friction of the packing rings when the valve is fitted with them. In some types of piston valves the live steam is admitted inside, between the heads, as shown in Fig. 71, and the exhaust passes out around the ends, but the same principle of balancing is retained as with the outside admission type, for the reason that the pressure is applied between the ends of the valve instead of on the outside as with the other type. The sketches here given do not show the valves in their true proportions, being merely used to illustrate the principle

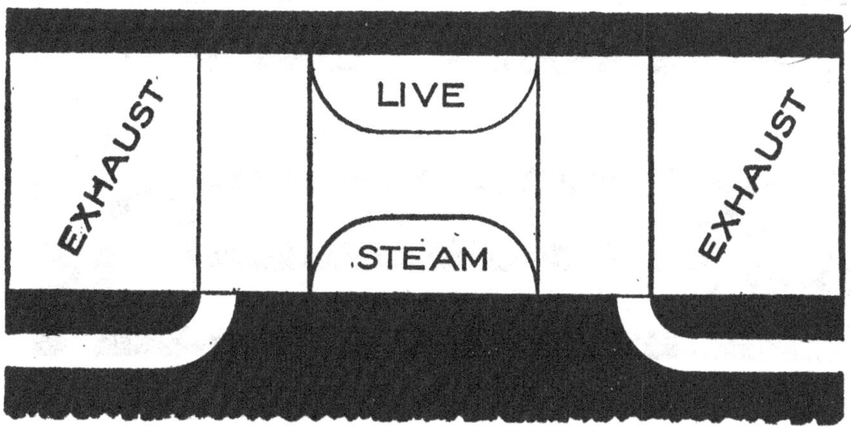

FIGURE 71

upon which the piston valve works. In practice the valve is made as long as possible, in order that the ports leading to the cylinder may be shortened to the minimum.

Another type of piston valve is shown in Fig. 72. This valve is made hollow for lightness and has packing rings at each end to prevent the steam from passing into the ports until at the proper moment. The edges of these packing rings control the admission of steam to the ports in the same manner as do the edges of the D valve, and when the valve is one of outside admission it is set in the same manner as the D valve is. But if admission is from the inside, as shown in Fig. 73, the movement of the valve is reversed, as is the method of setting also. As it is very essential that the packing rings at each end of a piston valve be steam-tight, a certain element of friction is introduced in this manner. In the larger number of cases where piston valves are used, central or inside admission is the rule, a great advantage of this type over outside admission valves being that the larger portion of the cooling surface of the valve chamber is reserved for

FIGURE 72

the exhaust steam. Another advantage is that of having only exhaust pressure against which to pack the valve rods, and make the joints for the heads of the valve chamber.

In taking charge of an engine having piston valves, an engineer should always first "look her over" and note the positions of the eccentrics with relation to the crank pin. He should also take a look at the rocker shaft if there is one. He will then be able to satisfy himself as to whether the valves have outside or inside admission, a very important thing to know in case anything should happen out on the road that necessitated resetting of one or both of the valves to enable him to bring his engine home. As before stated, the movement of a piston valve having outside admission is precisely the same as that of a D slide valve, but it is well to note the fact that while the great majority of engines fitted with D slide valves have indirect valve gear, still there are some in which the motion is direct. For the guidance of the engineer in such cases, the following four simple rules are here given.

Rule 1. If the eccentrics and crank pin are together, that is, on the same side of the driving shaft, and there

FIGURE 73

is a rocker arm that reverses the motion, the valve has outside admission, indirect.

Rule 2. If the eccentrics and crank pin are together and there is no rocker arm, but direct motion, the valve has inside admission, direct.

Rule 3. If the eccentrics and crank pin are on opposite sides of the driving shaft, and there is a rocker arm to reverse the motion, the valve has inside admission, indirect.

Rule 4. If the eccentrics and crank pin are on opposite sides of the shaft, and there is no rocker arm to reverse the motion, the valve has outside admission, direct.

There are, in fact, four possible combinations to deal with in the setting of locomotive piston valves, the first of which is the outside admission indirect connected valve, receiving its motion through the medium of the familiar rocker shaft, with one arm up and the other arm down. Second, inside admission direct, in which both arms of the rocker extend either up or down, and the forward motion of the eccentric rod produces a like forward motion of the valve. In these two combinations the eccentrics and crank pin are on the same side of the shaft. Suppose the crank pin to be on the dead center, then lines drawn from the center of the shaft through the heavy portions of the eccentrics would be approximately in the same position as would the hands of a watch indicating five minutes to seven o'clock, assuming the crank pins to be at 9 o'clock (see Fig. 74). Third, outside admission direct, in which the rocker arms do not reverse the motion of the eccentric blade; and, fourth, inside admission indirect, in which the motion is reversed by the rocker arms in the same manner as in combination one.

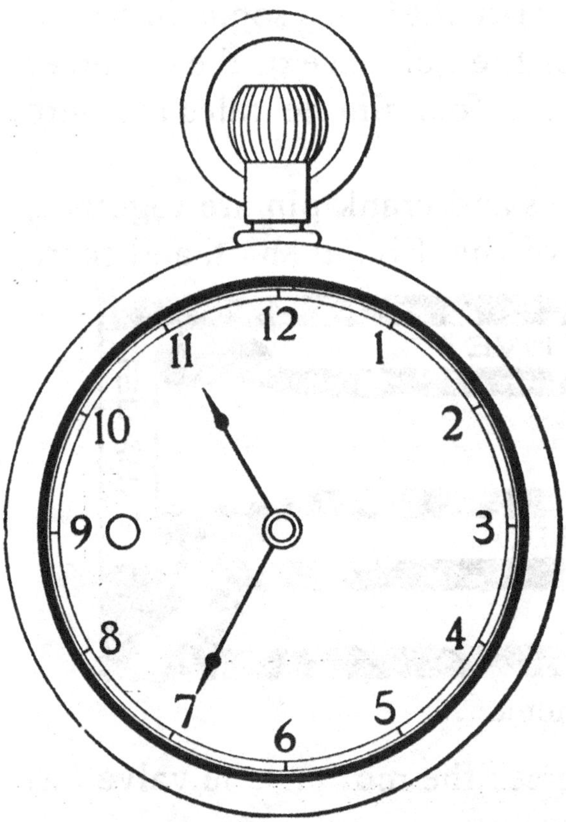

FIGURE 74

These two latter combinations may be termed the

p.m. setting, for the reason that lines drawn through the center of the shaft and the heavy portions of the eccentrics would occupy positions similar to the hands of a watch indicating five minutes past five, with the crank pin at nine o'clock (see Fig. 75), while the setting illustrated by Fig. 74 may be termed the a.m. setting, and as the careful engineer always has his watch with him, the following table may be of service:

Outside admission, indirect—5 min. to 7 a.m.

Outside admission, direct—5 min. past 5 p.m.

Inside admission, direct—5 min. to 7 a.m.

Inside admission, indirect—5 min. past 5 p.m.

Figure 75

A good rule to remember in setting piston valves is this: If motion is imparted to the valve on the a. m. plan as described, and it is desired to increase the lead, the valve must be moved towards the crank pin, but if the p. m. plan governs the motion it will be necessary to move the valve away from the crank pin to increase the lead.

The American Balanced Valve Co., of Jersey Shore, Penn., are the makers of a new type of piston valve, which they term "The American Semi-plug Piston

Valve." This valve, a description of which is here given, has performed very efficient service since its introduction, and it appears destined to occupy a prominent position in locomotive work in the future.

Referring to Fig. 76, an internal admission valve is shown. The inner sides of the two snap rings, 1-1, are beveled. The outer sides of the snap rings are straight and fit against the straight walls of the valve spool. Against the beveled sides of the snap rings, solid, uncut, non-expansible wall rings, 2-2, fit. Their

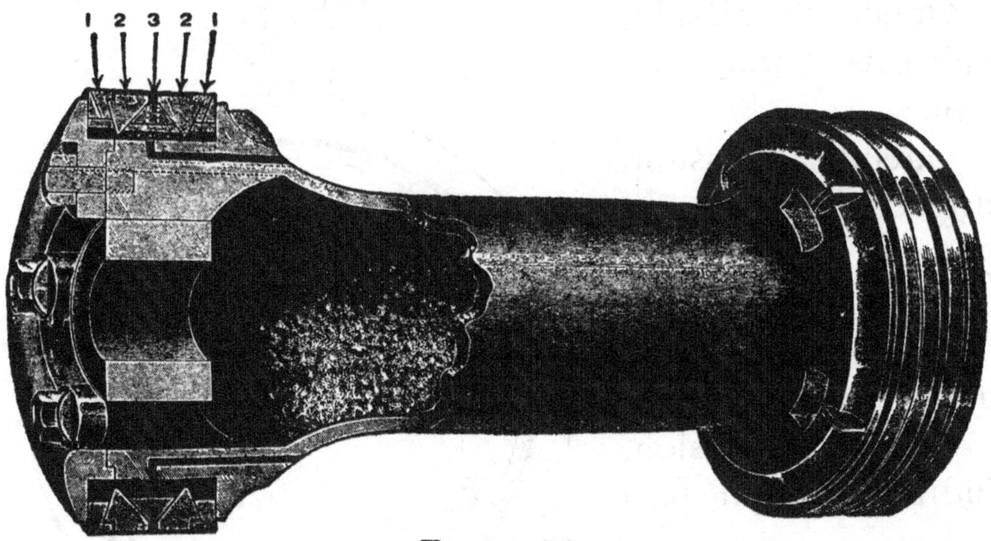

FIGURE 76

inner sides are beveled at a greater degree of angle than their outer sides, which fit the snap ring.

In between the two wall rings is placed a central double tapered snap ring, 3. This ring is properly lapped, and is put in under tension, thus holding the wall rings apart, putting a slight grip on the snap rings laterally. Thus applied, the action is as follows: When steam is admitted to the steam chest, or central portion of the valve, it passes through openings in the spool to the space beneath all of the rings, and acts upon the central wedge ring direct, giving it a lead of the snap rings in action, and forcing the wall rings against the sides of the snap rings, so that prevention of their excessive expansion is positive. The snap rings are thus expanded against the casing just enough to make steam-tight contact, and the central ring grips them there, and they are prevented from further expansion. This is demonstrated by withdrawing the valve from the valve chamber while under steam until the first ring in the spool is entirely out of the cylinder, when no increase in the diameter of the snap ring can be observed. It can then be pushed back into the cylinder again. It will readily be understood how easy it is to prevent further expansion of the snap ring by the pressure underneath it, when the degree of angle of the bevel on the inside of the snap ring is considered. By making this degree greater, the power of the central wedge ring would be sufficient to decrease the diameter of the snap ring, closing it away from the valve chamber. Therefore it appears that this valve has all the advantages of the plug valve, without the drawbacks of the plug valve, and it has all the advantages of the snap ring valve, without the drawbacks of the snap ring valve, because it is practically a plug that does expand and take care of itself, not only for the difference in contraction and expansion, but also for wear; yet the plug is not so rigid as to knock a cylinder head out before relieving the water from the cylinder, and yet it is absolutely adjusted to the diameter of the casing at all times, and is held there and allowed to get no larger during its work under pressure. The rings are so lapped that they are steam-tight from all directions, and the bevel lap joint maintains unbroken steam and exhaust lines at the edge of the ring.

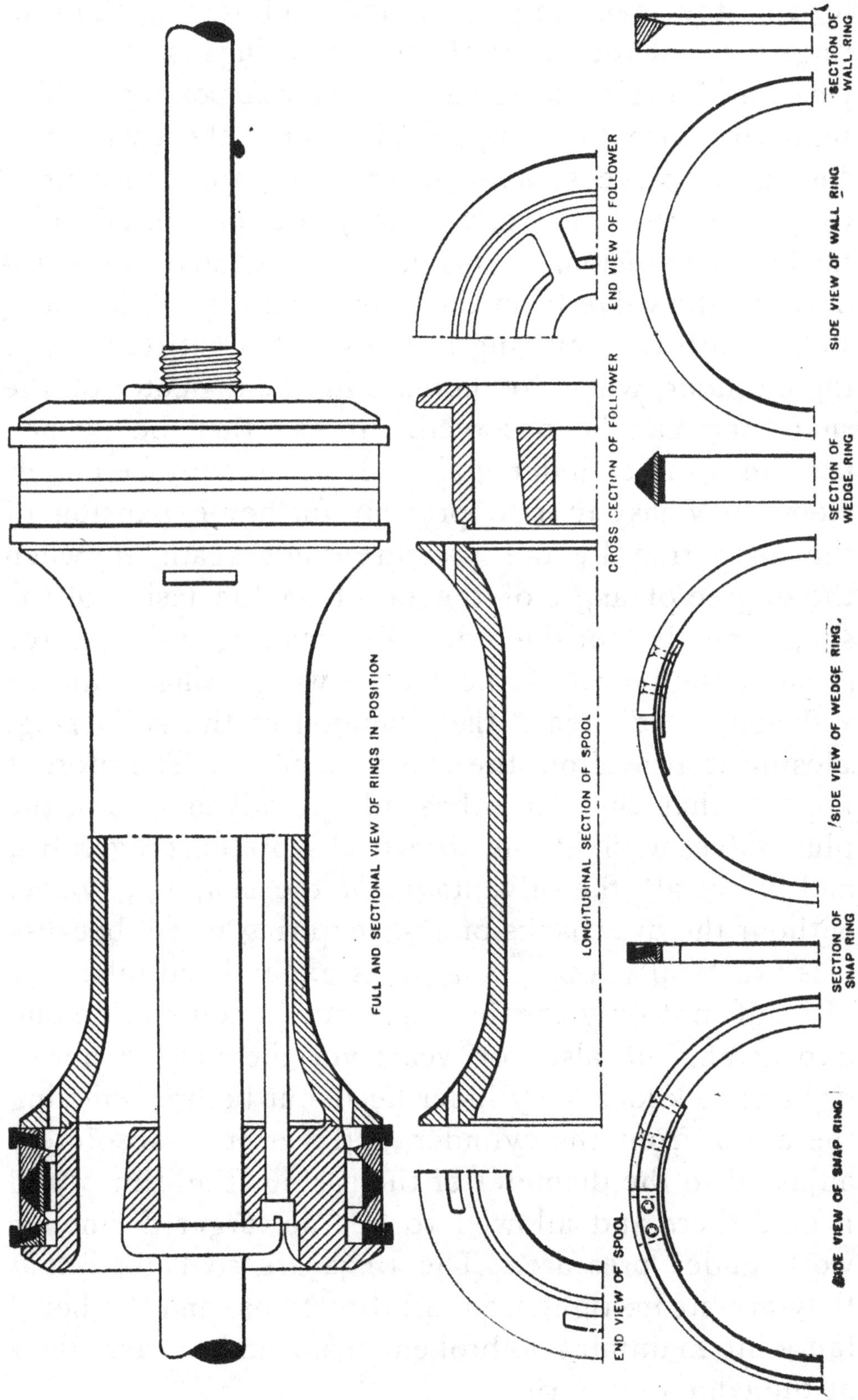

FIGURE 77 AMERICAN SEMI-PLUG PISTON VALVE

Fig. 77 further illustrates the construction of this valve, giving end and sectional views of the different parts. A common defect of snap ring piston valves is that the steam pressure gets under the rings, and expands them against the casing with the full force of the chest pressure, thus causing excessive friction, while at the same time the cage is worn unevenly by the valve working at short cut-off and over the ports. Under such conditions, steam-tight joints soon become leaky, and the leakage rapidly increases as the wearing goes on.

A piston valve, in order to give efficient steam-tight and durable service, should automaticaly regulate the frictional contact of the rings against the cage, and keep the cage perfectly true. It is claimed by the manufacturers of the American semi-plug piston valve that it meets these requirements, and the claim is substantiated by the record of an engine on the Buffalo and Susquehanna Railroad, which was fitted up with a set of these valves in June, 1901, and was in continual service up to April, 1904, or a little over two years and nine months, excepting when the engine was in the shop for necessary repairs, but during this time no repairs of any nature were required on the valves. No perceptible wear was detected, either of the casing or the rings, when the valves were removed for the purpose of exhibiting them at the St. Louis Exposition in 1904. It is also claimed that this valve does not require relief valves, by-pass valves, nor pop valves, and that it is handled the same as a slide valve, and drifts freely.

Many locomotives are equipped with piston valves of different types, but the internal admission valve appears to be the favorite. The form of piston valve used by the Baldwin Locomotive Co. on their Vauclain engine will be fully described in the section on compound locomotives.

One of the advantages the piston valve possesses over other forms of slide valves is that it may be made long enough to bring the two faces or working edges near the ends of the cylinder, thus greatly reducing the clearance between the valve face and the piston.

The term balanced valve, as used in this connection with reference to locomotive practice, is meant to include all balanced valves except those of the piston type. As stated at the beginning of this chapter, there have been many different kinds of balanced valves applied experimentally to the locomotive, by inventors in their efforts to reduce the friction between the face of the valve and its seat. It is stated upon good authority that up to January, 1904, there had been 573 patents issued to those who had made attempts to perfect the slide valve, but in the great majority of these cases failure has been written up against them. A few of the more meritorious of these will be described and illustrated.

The Jack Wilson High Pressure Valve is manufactured by the American Balance Valve Co., of Jersey Shore, Penn., and the following description of it is supplied by the makers, with the exception of a few minor changes in the text.

Valve. The valve, Fig. 78, is similar to the "gridiron" valve, it having two faces; one face operates on the valve seat proper (on the cylinder) and the other face operates against the face of the balance plate. Both faces of the valve are the same, and it has no crown, but is open throughout. The face of the balance plate, against which the top or back face of the valve operates, being an exact duplicate of the cylinder valve seat and set in alignment therewith, whatever conditions exist on one face of the valve must also exist on the other face. The walls of the valve are provided with ports, which pass from face to face of the valve. These ports are functional, and their length and width depend upon whether the valve is to be double or single acting; that is, whether or not double admission and double exhaust openings are desired. The valve under consideration here is of the double acting type.

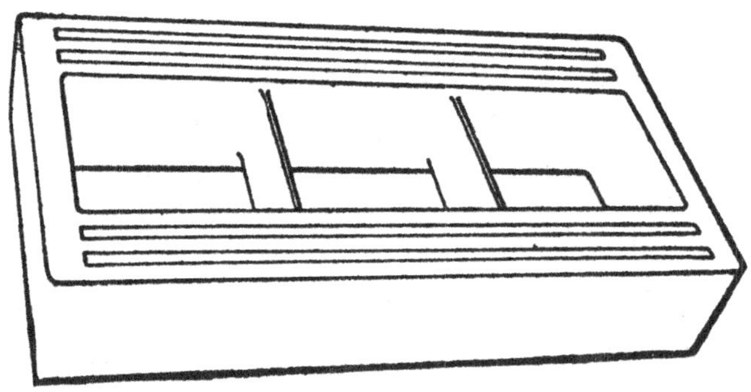

FIGURE 78

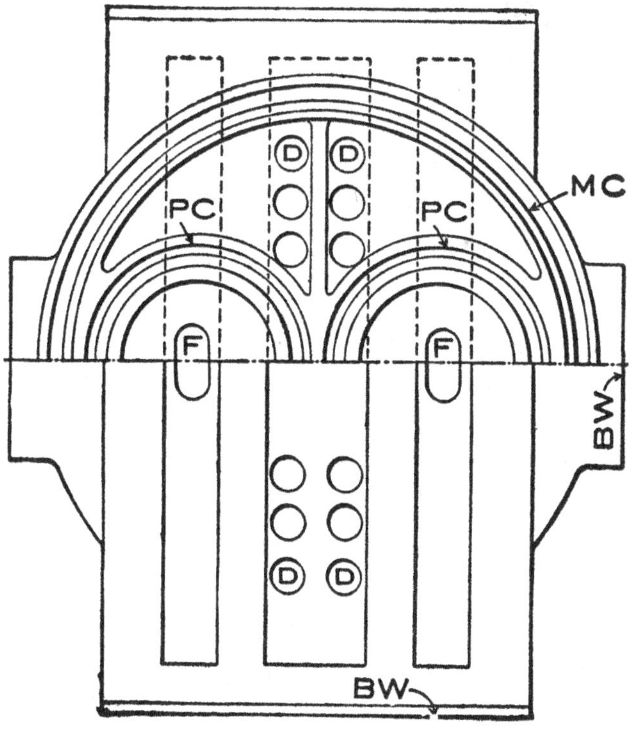

FIGURE 79

Balance Plate. The Balance Plate, Fig. 79, contains the balancing cones MC and PC (main cone and port cone), and two centering ring grooves which register with like grooves in pressure plate. It also supplies the means for double admission and double exhaust openings by admitting and exhausting steam at the face of the plate at top of valve simultaneously with admission and exhaust at valve face and cylinder valve seat.

The face of the balance plate, Fig. 80, is an exact duplicate of the cylinder valve seat and forms a second valve seat against which the valve operates in unison with its operation on the cylinder valve seat, the second seat being held by means of the centering rings CR, Fig. 81, in exact alignment with the valve seat proper. The back or opposite side of the balance plate, Fig. 84, contains the following cones: one large or main cone (MC) and two small or port cones (PC) on the interior of the main cone, and on which the packing rings are placed, which forms the balancing feature to the valve, and the centering ring cones. The balance plate is provided with wings (BW) which fit 1-16 in. loose into the wings of the pressure plate (or into the steam chest itself), preventing excessive movement of plate. Taper or beveled packing rings set on the cones form joints against the pressure plate.

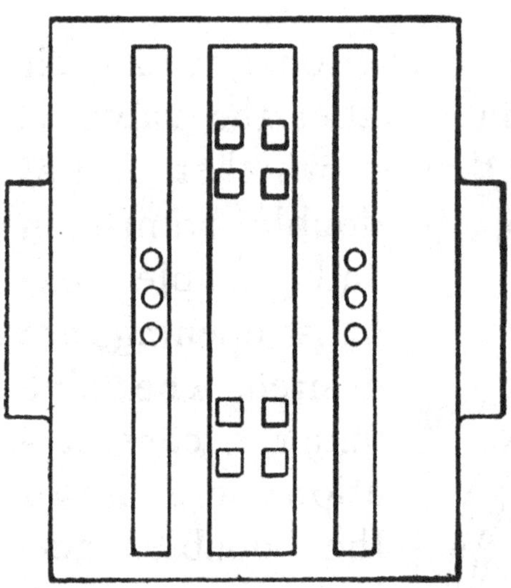

FIGURE 80

Pressure Plate. The pressure plate is made as a part of or separate from the chest cover. In the type of valve here referred to, the pressure plate is made separate (see Figs. 82, 83 and 84) and is provided with wings (W) which are machined to fit snugly into the steam chest; the chest being first centered with the valve seat by fitting over lugs on the cylinder, or by

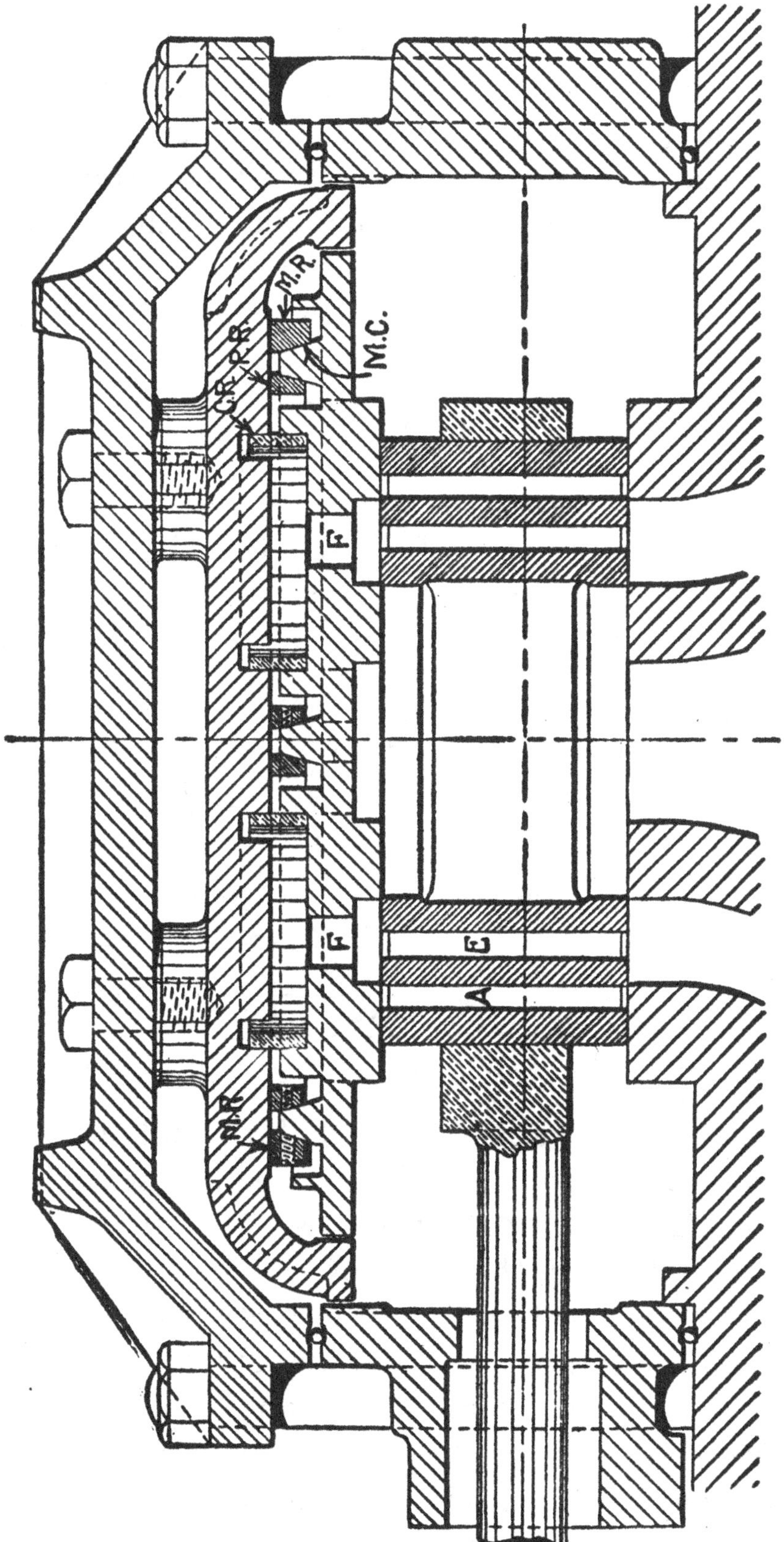

FIGURE 81

dowel pin, and machined at the top to receive wings (W) of the pressure plate. Into the face of the pressure plate two grooves are cut with either straight or taper walls and which register correctly with the corresponding grooves in the balance plate; these are called centering ring grooves and into them two centering rings (CR) are placed, slightly under tension.

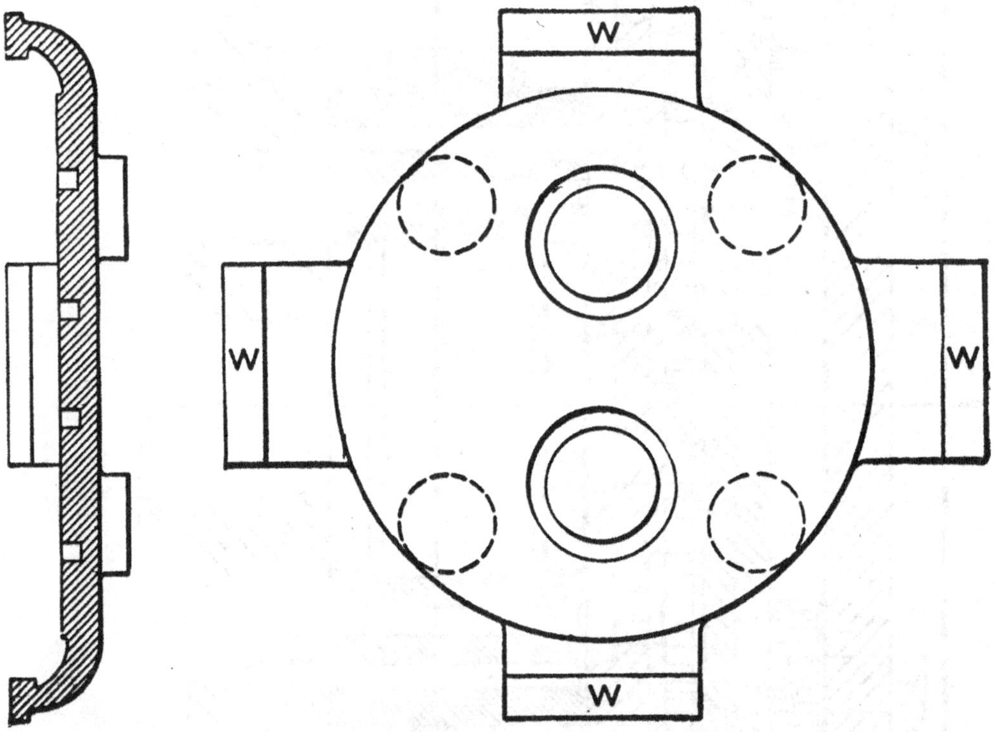

FIGURE 82

Under normal conditions these steel rings hold the balance plate in alignment with the valve seat, but under abnormal conditions, such as dry valves, the strain will be taken by the wing of the balance plate against the wing of the pressure plate, preventing excessive contraction of the centering rings. Against the face of the pressure plate the balancing rings form steam joints.

Balancing Feature. Having mentioned the three principal parts composing this valve, it is now in order to consider the balancing feature, which is of great importance, as it successfully protects the valve under the highest pressures. In considering the principle upon which the valve is balanced it is necessary to get clearly fixed in the mind the fact that the balanced area of the valve is changeable and that the change takes place automatically, so as to correspond with the changed condition of the valve on its seat at different points of its travel.

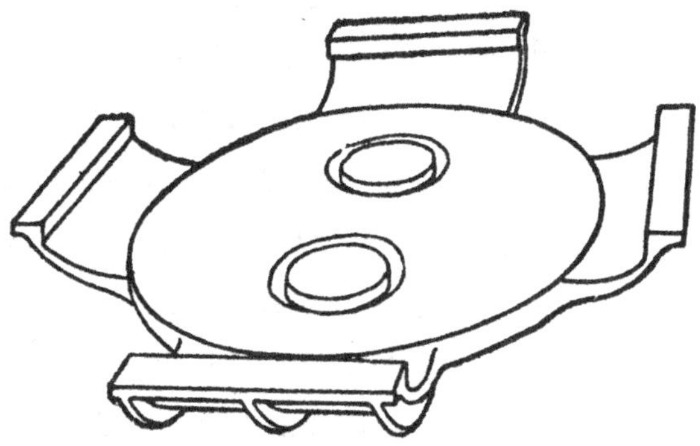

FIGURE 83

Referring to assembled cross sectional view, Fig. 81, the valve is seen in central position on the seat and the upper seat or face of balance plate in position corresponding with the valve seat. The steam chest is centered by machined faces fitting over machined lugs on the cylinder; on old power, dowel pins are used. The chest has finished strips at top into which the finished ends of the wings of the pressure plate (W) fit snugly, thus insuring the central position of the pressure plate over the valve seat. The finished wings of the balance plate (BW) fit 1-16 in. loose between inside faces of the wings of the pressure plate, but the

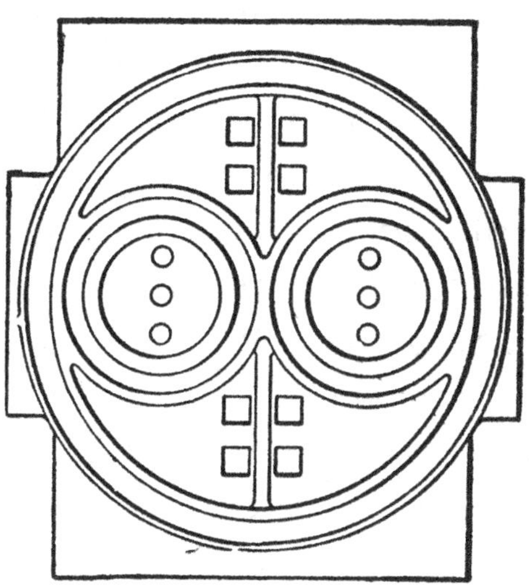

FIGURE 84

balance plate is held perfectly central by two steel centering rings (CR). The tops of the cones on the balance plate are ⅛ inch from the face of the pressure plate, allowing the balance plate to lift ⅛ inch off from the valve, which affords perfect relief to the cylinder while the engine is drifting and for the relief of water from the cylinder. This ⅛-inch clearance in height adjustment must be maintained. The main balancing ring (MR) is made the proper diameter to balance the valve as great as possible while in its central (or heaviest) position, there being just sufficient area left on to insure the balance plate being held steam-tight on the valve. The interior of the main ring is open to the atmosphere through the holes D, which lead to the exhaust cavity of the valve.

The valve is thus balanced so that it will move perfectly easy in its heaviest position, but conditions are changed by the opening of a steam port (and at instant of cut-off. See Fig. 85), at which time the ordinary slide valve is subjected to the upward pressure of the steam in the cylinder port, and if properly balanced in central position would, at this position, be thrown off its seat, but in this valve the port pressure (whatever it may be) has free access to both sides of the valve by reason of the passages through the valve to the port in the face of the balance plate which corresponds with the cylinder port; therefore the pressure in the port has no effect whatever upon the valve, it being on both sides of the valve face in equal area, and pressure, is, therefore, equalized so far as the valve is concerned, but the pressure in the port of the balance plate would lift the plate off from its seat on the valve if it was not also equalized, or annulled; therefore a port ring, PR, of proper area to balance this pressure,

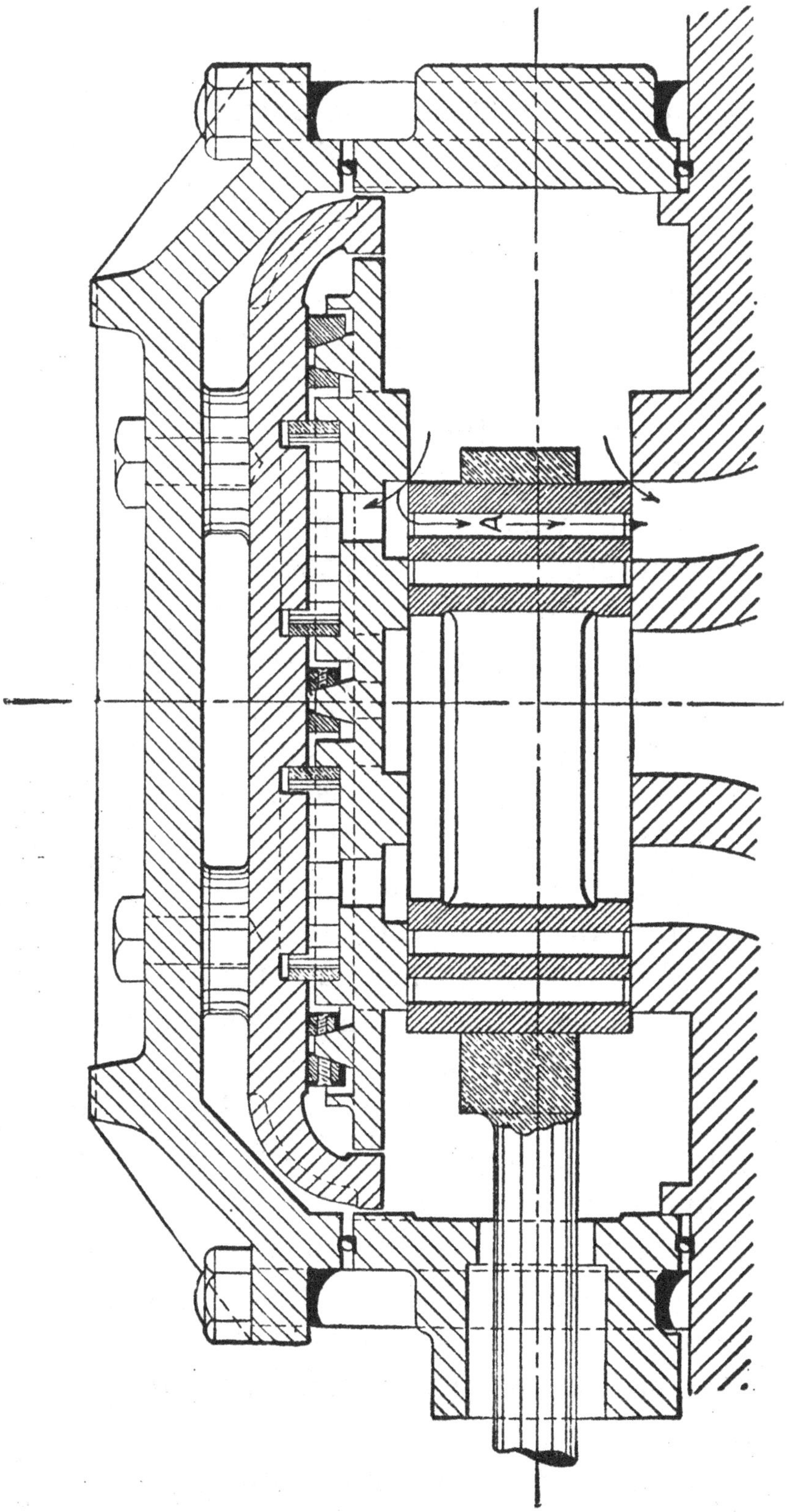

FIGURE 85

is placed over each port in the inside of the main ring on the top of the balance plate and is open to the port through passage F, Fig. 81, so that a pressure equal to that in the steam port is always on both sides of the balance plate, as well as on both sides of the valve, and the port pressure is rendered inoperative on the valve or on the balance plate. Communication from the cylinder port, through the valve and through the balance plate to the interior of the port ring, P R, cannot be shut off at any time, but is maintained throughout the travel of the valve. Therefore the same pressure that is in the port at any given time is also on both sides of valve and pressure plate in the same area, and the port pressure is, therefore, not considered in figuring the main balance for the valve.

There is another position of the valve during its stroke where the slide valve is subjected to an upward pressure, or pressure against its face, which tends to lift it from its seat; that position is at over-travel of the valve face over the valve seat; this position is shown in Fig. 86, but in this valve it will be observed that the top face or back of the valve travels out from under the seat of the balance plate exactly to the same extent that it over-travels the cylinder seat, and pressure is, therefore, equal on both sides of that portion of the valve that is over the seat at any point of travel. With the main ring balancing the valve fully in its central or heaviest position, the port ring balancing the port pressure, and over-travel of the valve on its seat being equalized by equal exposure at top and bottom, it will be clear that the valve is fully balanced in all positions of stroke, and is, therefore, available for high pressures.

The double admission of steam to the cylinder and

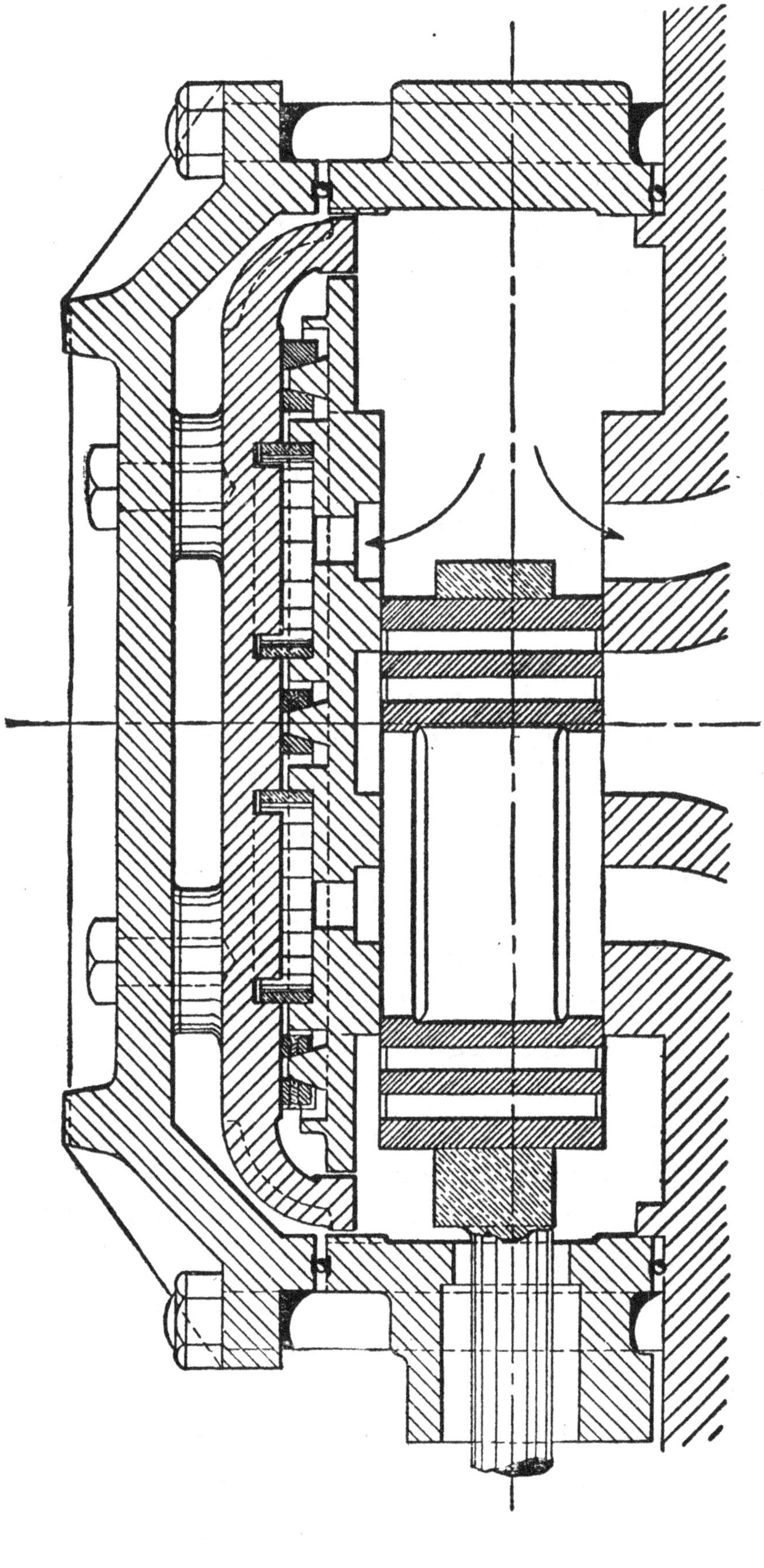

FIGURE 86

the double opening for exhaust of same are made clear in Figs. 85 and 87, which show the valve at point of admission and point of exhaust respectively. Referring to Fig. 85, the valve is admitting steam to the cylinder port direct, and at the same time is admitting steam to the port (pocket port) in the balance plate and thence by way of passage A through the valve into the cylinder, thus securing double admission openings. Note direction of arrows.

Referring to Fig. 87, the valve is opening for exhaust and the steam leaves the cylinder at the face of the valve at cylinder seat and also by way of passages E through the valve into the port (pocket port) in the balance plate, and out at the face of the valve, thus securing the double opening for exhaust, which has always been considered a feature much to be desired in the locomotive valve. Owing to the fact that the travel of the valve over its seat is equalized, it is possible to sc proportion the width of the valve seats that the valve travels to the edge of, or slightly over, the seat when the engine is worked at the shortest possible cut-off, and the valve must, therefore, make a full stroke across the seat or "wipe the seat" at every revolution of the wheel, regardless of the cut-off; perfectly straight wear of the valve seat is the result.

In applying the valve to the engine it is important that the face of the balance plate, or upper valve seat, shall be in alignment with the cylinder seat in order to secure simultaneous action of the valve, at both faces, as previously explained; this is accomplished in various ways, one, a very positive and easy method, being shown here.

The height adjustment is ⅛-in. clearance for lift of valve or balance plate for relief of water from the

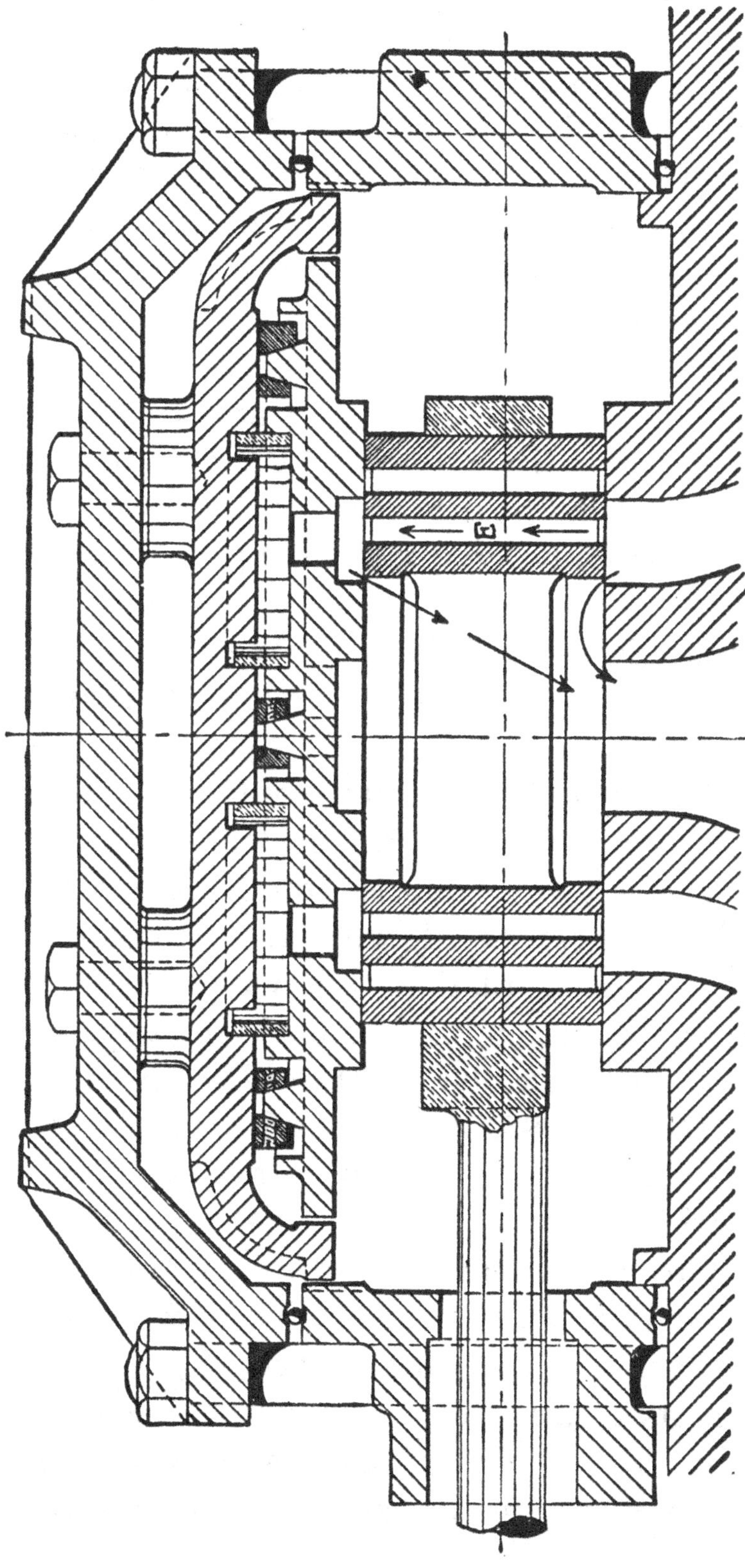

FIGURE 87

cylinder and to open direct communication from one side of the piston to the other for the free passage of air in drifting. In this connection, it should be observed that the balance plate will leave its seat on the valve while the valve remains on the valve seat, and that, while the balance plate is off its seat, direct communication from one cylinder port to the other is always maintained by reason of the ports AE through the valve; this affords the most perfect air relief for drifting.

The packing rings remain stationary and are, therefore, subject to practically no wear; they afford full automatic adjustment to position and for wear of valve faces and are free from danger of breakage or derangement.

The Richardson Balanced Valve. This form of balanced slide valve, together with the Allen-Richardson balanced slide valve, is manufactured by H. G. Hammett of Troy, New York, and is largely used on locomotives. Figs. 88 and 89 represent transverse and longitudinal sections through the center of an ordinary locomotive steam chest fitted with the Richardson valve. Fig. 90 shows a plan of the valve, and Fig. 91 is an elevation of one end of the packing strips and spring, the only alteration being the addition of the balance plate, PP, Fig. 88, and the substitution of a valve adapted to receive the packing strips S, S, S, S.

It will be noticed in this instance that the balance plate is bolted to the cover of the steam chest, but these may be cast in a single piece. The four sections of packing enclose a rectangular space, R, Fig. 90, which equals in its area the total amount of valve surface which is to be relieved of excess pressure, the packing strips preventing the steam from entering this space, and the small hole X, communicating with the exhaust cavity in the valve, relieves space R from any possible accumulation of pressure.

The four packing strips consist of two longer ones, which are simply rectangular pieces of cast iron, while the two shorter ones, Fig. 91, have gib-shaped ends to retain them in their proper position. Beneath each packing strip a light elliptic spring, shown in Fig. 91,

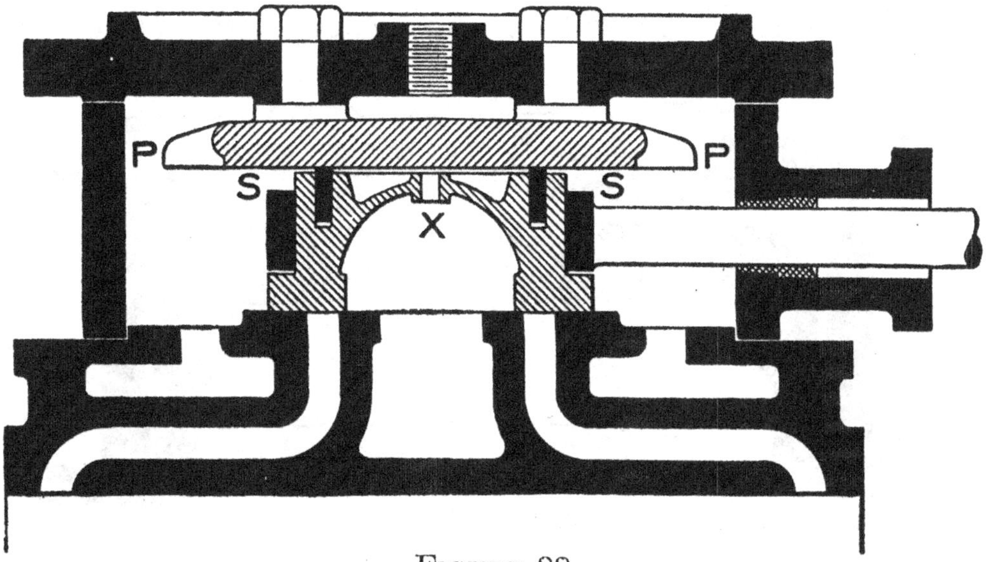

FIGURE 88

is placed which holds these strips in position against the balance plate when steam is shut off.

In operation these different sections maintain a steam-tight contact, by a direct steam pressure, with the balance plate and with the inner surfaces of the grooves provided to receive them, the joint being secured by the abutting of the ends of the two longer sections against the inner surfaces of the gibbed sections at the four corners.

The Allen-Richardson Balanced Slide Valve. The Allen valve is designed to at least partly prevent the wire-drawing of the steam, when high speeds are maintained with the valve cutting off early in the stroke.

In the Allen valve, an additional passage for the inlet of steam is furnished, as will be clearly seen by referring to Figs. 92 and 93. These are transverse and longitudinal sections through the valve and steam chest, and it will be noticed that, when the steam port is open one-half inch in the ordinary manner, the port of the cored passage is also open to a like extent on the other side of the valve; consequently the effective

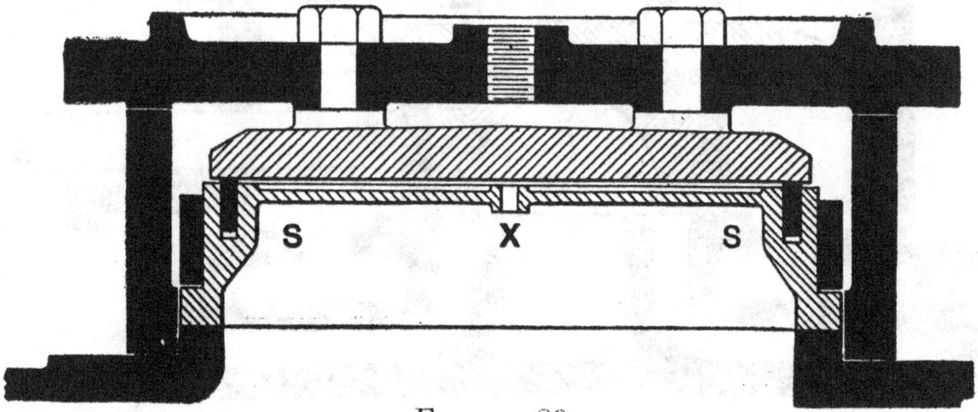

FIGURE 89

area of the steam port is doubled, and is thus the actual equivalent of a single port with a one-inch opening.

The wire-drawing incident to running at high speeds with the valve cutting off early in the stroke, is thus greatly diminished, with a resultant economy of steam and fuel. A reduction of wire-drawing carries with it a higher average pressure on the piston when working at a similar cut-off; consequently the usual average pressure can be maintained with a shorter cut-off, resulting in an appreciable economy. While the unbalanced Allen valve, therefore, secures a better and more economical distribution of steam, its use entails certain disadvantages.

On the face of a slide valve, the area of bearing surface is never sufficient to secure its wearing well under a heavy steam pressure; and this wearing surface is yet further reduced in the Allen valve, owing to its internal steam ports. This internal passage actually divides the valve into two parts, and the steam pressure, acting on the outer part, springs and bends its working face below that of the internal or exhaust port of the valve. The available wearing face is consequently reduced to a space about one-half as wide as the out-

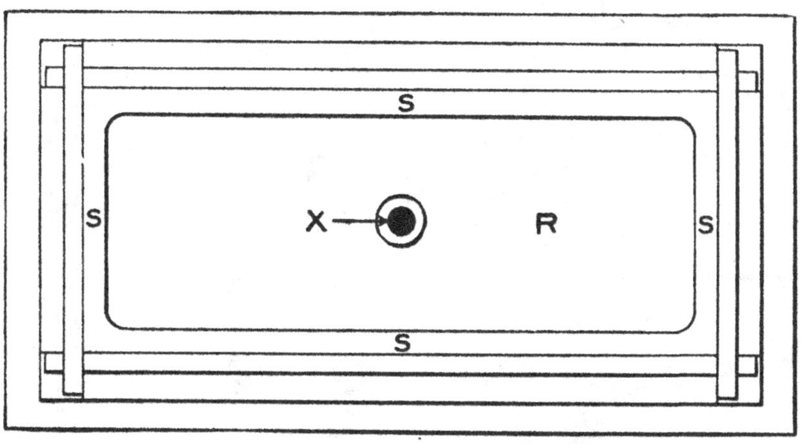

FIGURE 90

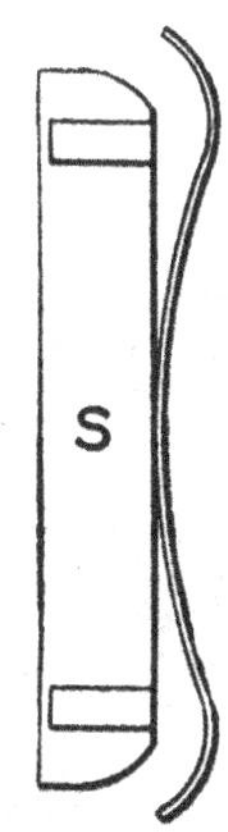

FIGURE 91

side lap of the valve, and this fully accounts for the rapid wearing of the unbalanced Allen valve, and for the trouble and expense of constantly refacing valves and seats, and the loss of the steam blown through leaky valves quite offsets the advantages gained by a reduction of wire-drawing.

These manifest disadvantages are entirely overcome by a proper balancing of the valve, which secures all of the advantages of the Richardson device, plus an increased steam economy resulting from using the Allen ports.

To secure the best possible results from the employment of the Allen balanced valve, its ports and bridges should exceed the full travel of the valve by at least one-eighth of an inch, and the radius of the link should always be as long as permissible to escape an excessive increase of lead when cutting off early in the stroke.

The Young Valve and Gear. During the past four years there has been brought to the front a valve which, while not a balanced valve in the ordinary acceptance of the term, as applied to locomotives,

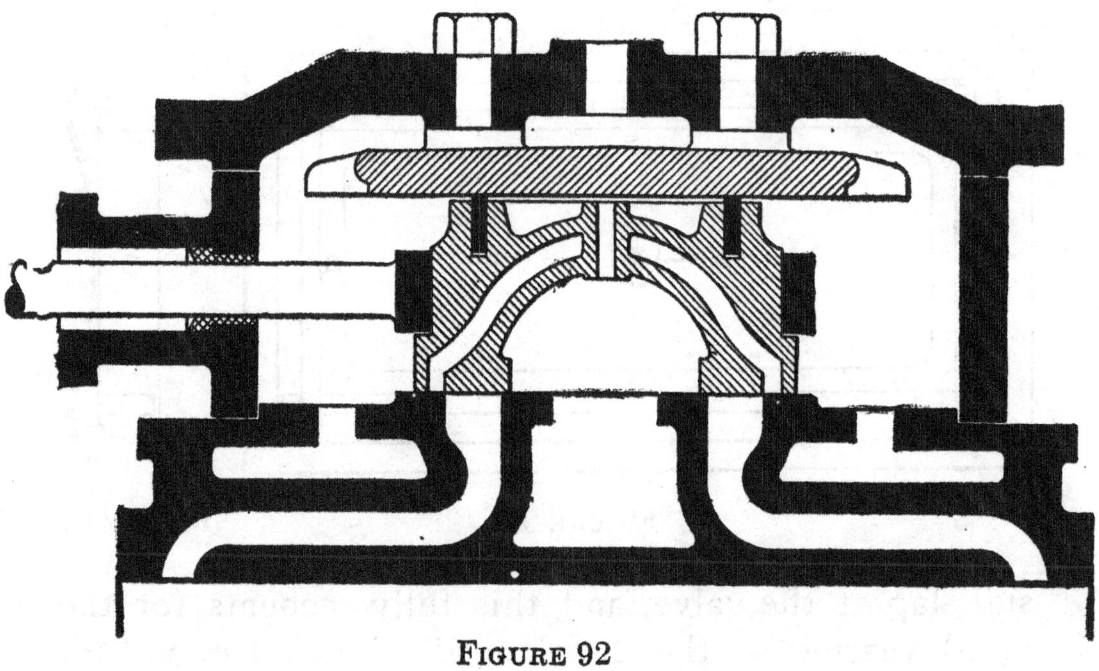

FIGURE 92

nevertheless gives or appears to give as good a distribution of the steam as either the balanced slide valve or the piston valve, while at the same time its operation is accomplished with a minimum of friction and strain on the valve gear. This is the Young valve and gear, the invention of Mr. O. W. Young of Chicago. The results obtained by the practical use of this valve on one of the engines of the Chicago & Northwestern Railway, especially during the past year, seem to warrant the conclusion that it has many meritorious features, and that a bright future lies before it.

A general idea of the construction of the valves, and the wrist plate by which they are operated, may be obtained by reference to Fig. 94, which is a sectional elevation showing the steam and exhaust ports, and a sectional view of the two valves, one for each end of the cylinder. The arrows clearly indicate the course of the steam in its passage into, and out of, the cylinder.

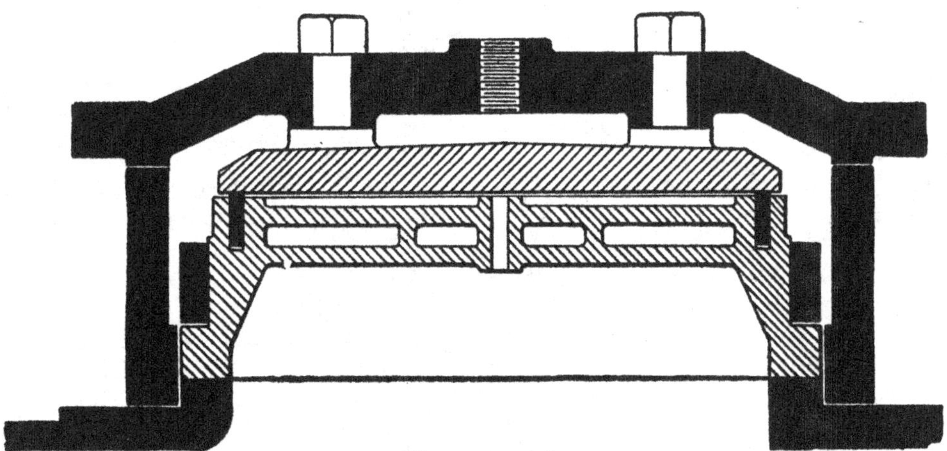

FIGURE 93

It is claimed for this system that irregularities in lead are corrected for the shorter points of cut-off, and the indicator diagrams shown in Figs. 96 and 97 certainly show an excellent steam distribution for all points of cut-off.

The author desires to say in this connection that he has seen the originals of these diagrams as taken from engine 1026 of the C. & N. W. Ry. and can vouch for their correctness.

The following brief description of this valve is furnished by the inventor:

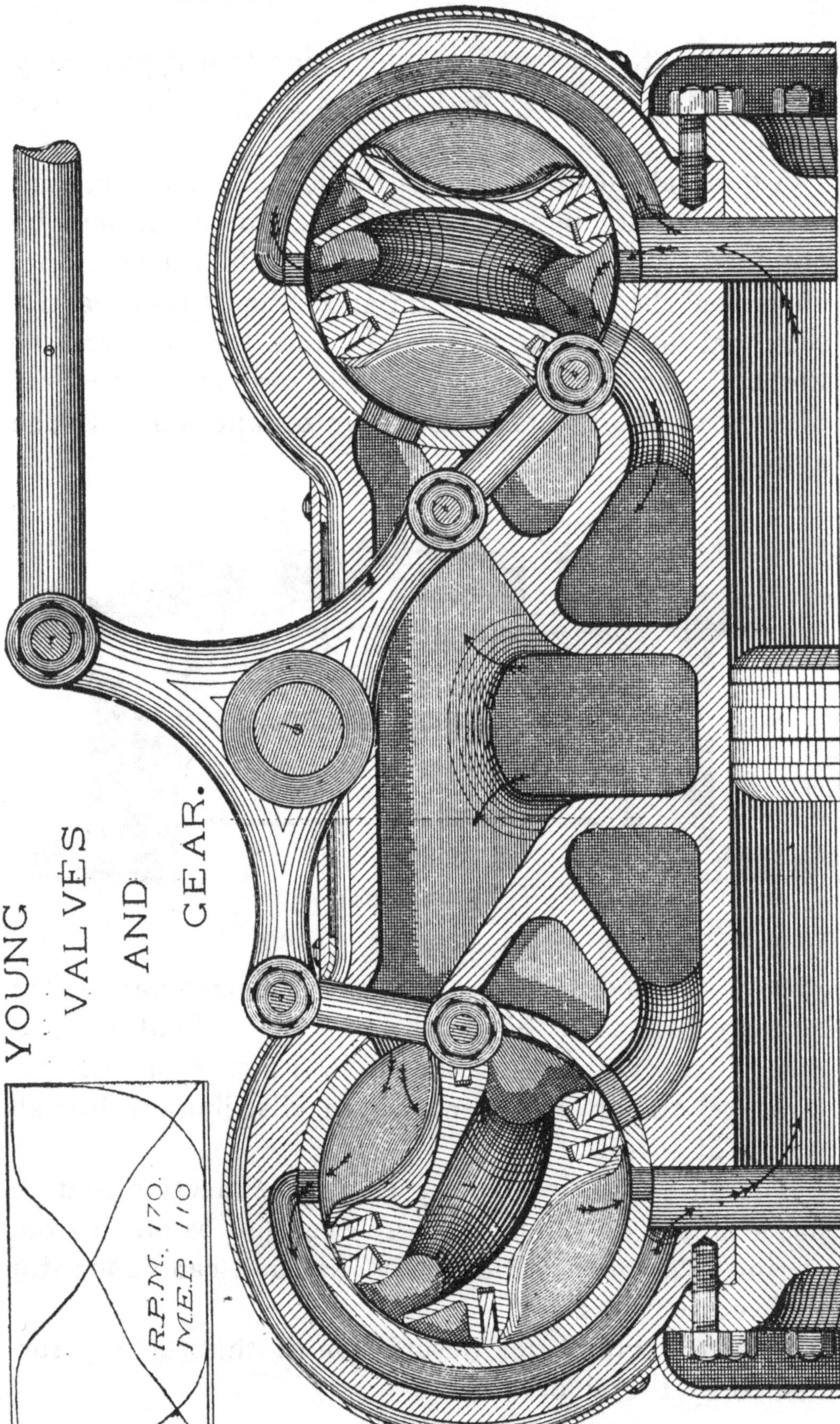

FIGURE 94

The Young valve and gear is an adaptation of the Corliss principle to suit requirements in locomotive practice, and consists of two valves for each cylinder, operating alternately as inlet and outlet and driven by the Corliss wrist motion, used in connection with the Stephenson link. An original device is provided for correcting the irregularities in lead, and either a constant or a slightly increased steam lead for the shorter cut-offs can be obtained, and an excessive preadmission of steam avoided. The exhaust lead, by this device, is caused to increase as the cut-off is shortened and permits an exhaust lap for long cut-offs, changing to exhaust clearance for a short cut-off, thus securing the maximum of power while starting (as shown by the straight back pressure line in the indicator cards) and sufficiently late compression to prevent the terminal pressure from exceeding the initial pressure even at very high speeds, and this is accomplished without the aid of by-pass or compression valves of any description.

The valves consist of a plurality of cast iron strips encompassing the exhaust cavity and partitioning the live from exhaust steam, and are each free to move towards and from their seat independent of each other; each following its individual path of travel and adjusting itself to any irregularities in the seat over which it moves, thus reducing leakage to a lower amount than is usually accomplished. The valve body or carrier is journaled at each end and its weight supported entirely clear of the valve seat, the only weight on the seat being that of the strips; the tendency, therefore, towards cutting, as compared with a heavy slide valve, is reduced to a very small percentage, and the necessity for liberal lubrication is obviated. The valve stems in their passage through the walls of the steam chest require no lubrication or packing. They will continue steam-tight and require no attention between shoppings in the way of taking up lost motion. Valve renewals are confined to the substitution of one or more strips.

The valve gear consists of the ordinary Stephenson link, eccentrics, and rocker-arm as far as the end of the valve stem, which is connected to the wrist plate.

From the wrist plate extend short hinged connecting rods to crank arms on the two rotative valve spindles.

The wrist plates are located between the cylinder saddles and the steam chests, and rotate on trunnioned bearings.

Fig. 95 shows a general plan and elevation of this system. The device for correcting the lead is operated in the following simple manner:

By reference to Fig. 95 it will be seen that a horizontal shaft extends across the back of the cylinder saddle, and that this shaft is fitted with two cranks that connect with the bearings of the wrist plates. From the center of this shaft a long connecting rod extends back to a short crank arm on the tumbling shaft. So that when the link is raised or lowered from the central position, the wrist plate is raised to regulate the lead. Experiments with this valve show that the best results are obtained by allowing about ⅛ in. more lead for the shorter cut-off. The valve chests are fitted with 10-in. bushings, within which the valves rotate. These bushings are made of soft cast iron, and the valves, as has already been explained, are fitted with cast iron packing strips.

The clearance in the cylinder of the first locomotive equipped with this device was 3 per cent and in the second engine it was 6 per cent Experience thus far

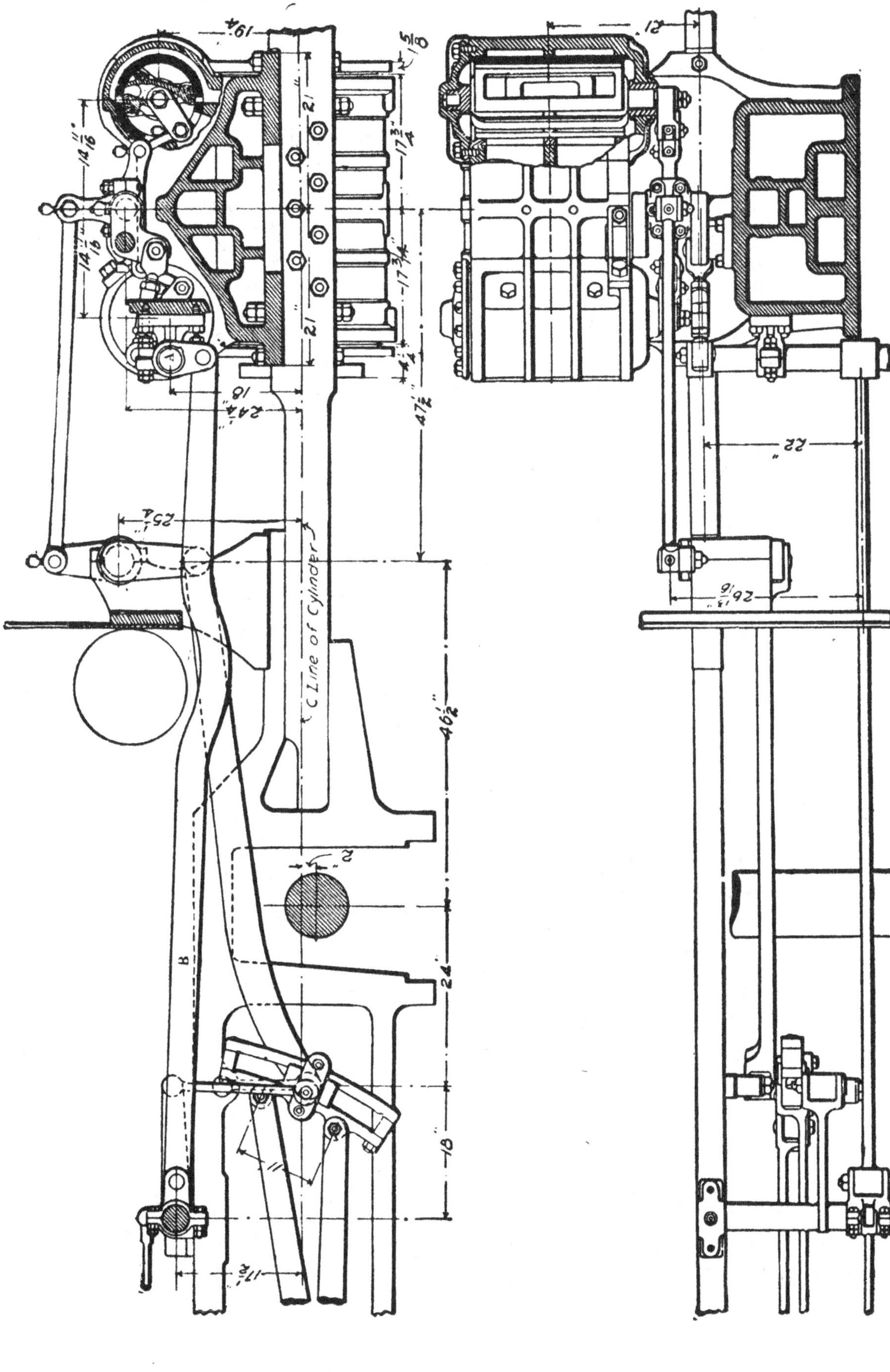

FIGURE 95

gained shows that 5 per cent would be the best figure.

Mr. Robert Quayle, superintendent of motive power

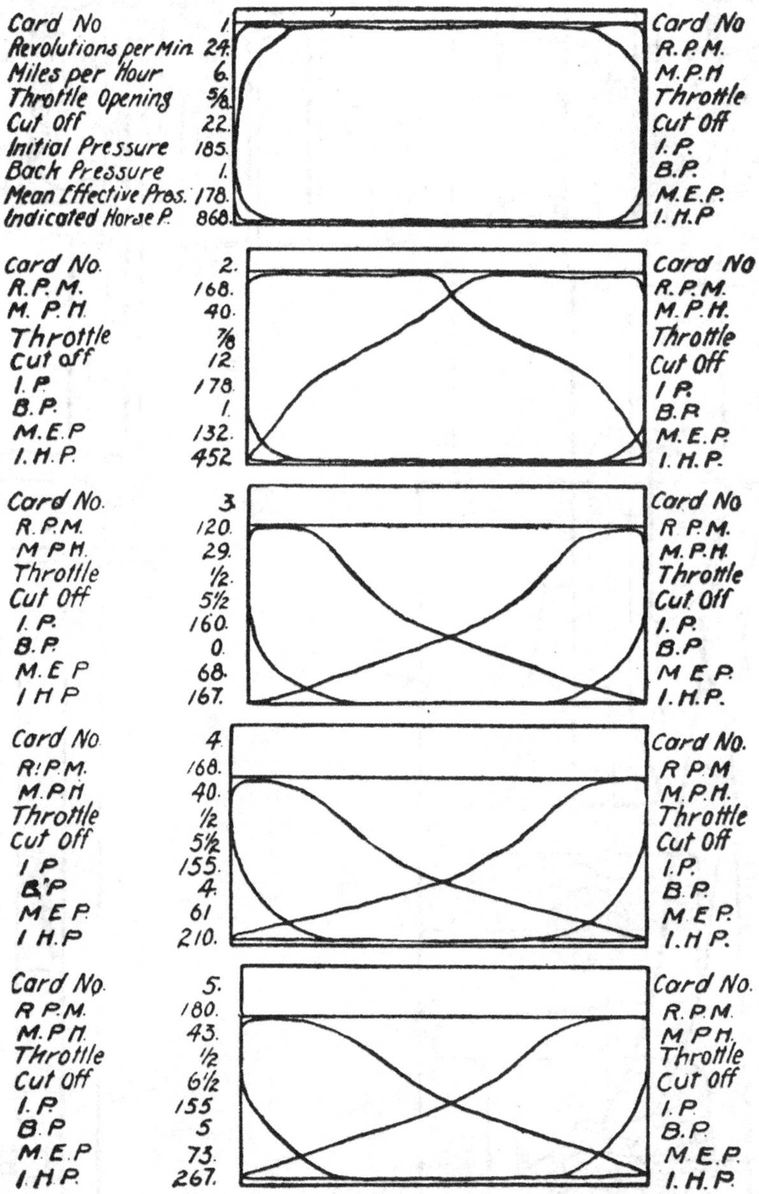

FIGURE 96

and machinery for the Chicago & Northwestern Railway Company, has kindly furnished the author with the following information regarding the testing and development of this interesting device on the Northwestern.

"The Young valve and gear has been developed on the C. & N. W. Ry. under the direct supervision of Mr. O. W. Young. There are at present on the Chicago & Northwestern Railway two locomotives equipped with the Young valve and gear, which is a system of rocking valves (two to each cylinder) which are operated by the usual eccentrics and links of the Stephenson motion. The construction of the valves requires an especial cylinder casting and therefore it cannot be used without a complete change.

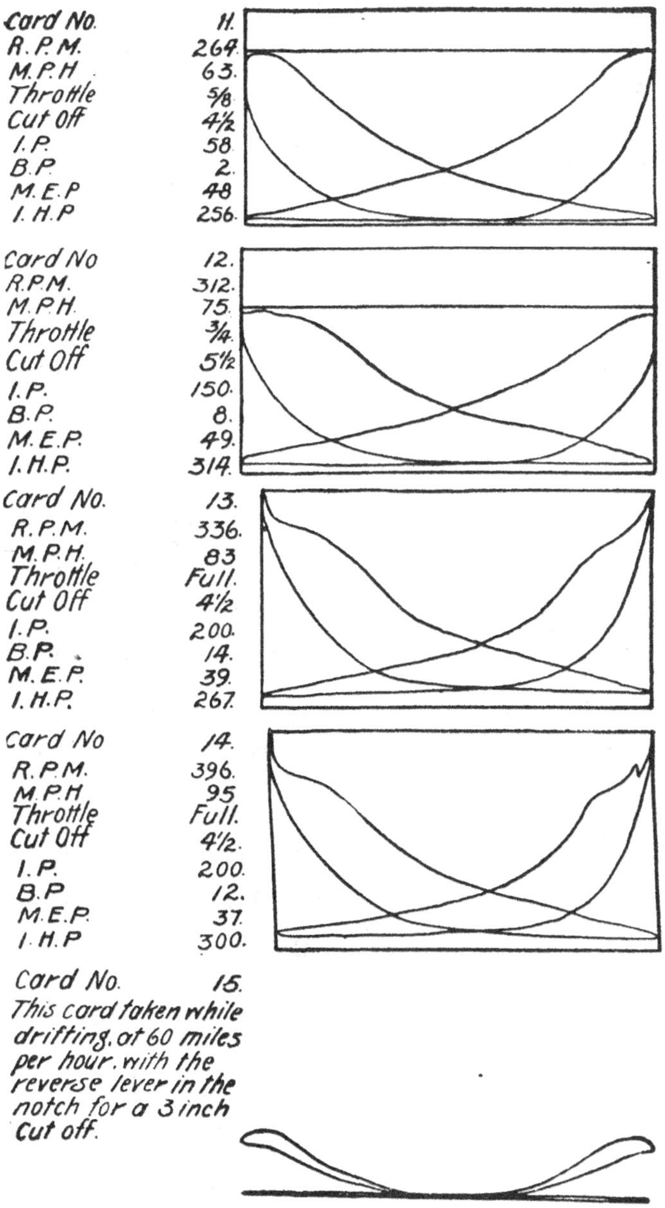

FIGURE 97

The actual cost of these cylinders, including the valves and changes in the motion, should not exceed thirty per cent more than the cost of cylinders, valves, chests, etc., for a D or piston valved locomotive. If, however, these cylinders were made standard to a road, I do not think they would cost more than $150 more per locomotive.

In June, 1901, the first engine was equipped, and, like all first attempts, there were certain details shown up which needed improvement. The general results with this engine justified a second trial, and in September, 1903, a set of cylinders with the special valves and their motion were applied to a 20 × 26″ Atlantic type (passenger) engine with 81″ over the tires and 91,000 lb. on the drivers. The experiments with this engine lasted some six or eight weeks, and in November, 1903, the engine was put regularly into service on the Galena division. The engine has been a "tramp" up to a very recent date; has had all kinds of service, all kinds of engineers handling her, and practically continuous service. It has so far made approximately 90,000 miles. The tires have not been turned, the eccentric straps have been closed once about $\frac{1}{16}$ in. each, there is no pound in the boxes and the tool marks are still on the motion pins. These results are especially interesting to the motive power official, demonstrating as they do that the wear and tear on the machinery is so remarkably less than the engine with the D or piston valves. The engine is always ready for service, the roundhouse foreman reporting that for his part of it, five of this type would easily equal seven of the piston valve engines. There is one run between Chicago and Clinton, with usually ten heavy cars, on which this engine is the only one that can make the time.

The train dispatchers know the value of this engine,

also, as they do not hesitate to rely on it to make up time or take an unusually heavy run. As a consequence the improvements shown by the indicator cards are not entirely realized in actual performance records. In a series of comparisons made by the indicator the water rate per indicator horse-power was reduced from 22.9 lb. to 19.3 lb. The indicator cards also show the cause for the slight wear on the machinery, as the cards are remarkably full, the expansion lines being clear and distinct at all points of cut-off. Most of the work in passenger service is done at less than 6 in. cut-off. On account of the high and full cards it is evident that the crank effort is uniform and higher than a slide valved engine. Besides causing less wear on the machinery, this gives a more even torque when starting and the consequent less slipping.

The engine is one which will bear thorough investigation. While our experiments have been made in passenger service, I consider that the performance in freight service will show even better results from both an operation and economical standpoint."

It should not be inferred that the Young valves are designed for light throttle conditions only. These engines, as will be seen by reference to Figs. 96 and 97, respond readily to a wide open throttle, and one diagram, No. 14, shown in Fig. 97, indicates a speed of 396 R. P. M. and 95 miles per hour, with full throttle and cutting off at four inches.

Fig. 98 shows a general view of engine 1026 of the C. & N. W., which engine is equipped with this valve.

Questions

262. Are there any other types of valves used on locomotives, besides the D slide valve?

FIGURE 98

263. Mention a few of them.

264. What is the principal objection to the D slide valve?

265. Is the piston valve a perfectly balanced valve?

266. What pressure tends to press the D valve against its seat?

267. Is there any pressure to counteract this?

268. What are the causes of friction in piston valves?

269. Are all piston valves outside admission valves?

270. Why are piston valves practically balanced?

271. Why are piston valves made as long as possible?

272. What controls the admission of steam to the ports of a piston valve engine?

273. How is an outside admission piston valve set?

274. How is an inside admission valve set?

275. What advantage results from using an inside admission piston valve?

276. Name another advantage in inside admission.

277. What is one of the first duties of an engineer taking charge of a piston valve engine?

278. Why should he do this?

279. Are all engines equipped with indirect valve gear?

280. Repeat four simple rules for the guidance of an engineer in the study of valve gear.

281. Mention four possible combinations that may have to be dealt with.

282. In the first two of these, what are the positions of the eccentrics with relation to the crank pin?

283. What time would the hands of your watch indicate to represent this setting?

284. What would be the positions of the eccentrics relative to the crank pin in the third and fourth combinations?

285. What time would your watch indicate in order to correspond with this setting?

286. What is a good rule to remember in setting piston valves?

287. What type of piston valve does the American Balance Valve Co. manufacture?

288. Describe in general terms the construction of this valve.

289. What force expands the packing rings of this valve?

290. What prevents excessive expansion of these rings?

291. What is a common defect of snap ring piston valves?

292. How may the valve cage be worn unevenly?

293. What should a piston valve do in order to give efficient service?

294. What type of piston valve appears to be the favorite with builders?

295. Mention another advantage possessed by the piston valve over other forms of slide valves.

296. What does the term balanced valve include in its definition?

297. Have there been very many types of balanced valves tested on locomotives?

298. By whom is the Jack Wilson high pressure valve manufactured?

299. Give a short description of this valve.

300. Is it single acting or double acting?

301. What is the function of the balance plate?

302. What is the object of the pressure plate?

303. Is this valve balanced at all points of its travel?

304. What advantage is gained by having the valve

make a full stroke across the seat at every revolution?

305. Describe the Allen-Richardson balanced valve.

306. How is the balance plate of the Richardson valve secured in place?

307. How is a steam-tight contact maintained between the upper surface of the valve and the balance plate?

308. How are these packing strips held in position?

309. Does the unbalanced Allen valve wear well?

310. What is the object of the internal passage in the Allen valve?

311. In what respect does the Young valve differ from the majority of valves as applied to locomotives?

312. What is claimed for this valve with regard to lead?

313. Describe in general terms the Young valve and gear.

314. Where are the wrist plates located in this system?

315. Describe the device for automatically regulating the lead.

316. What is the diameter of the bushing within which the valve rotates?

CHAPTER VII

THE INDICATOR

The Indicator. One of the greatest aids to the economical operation of the steam engine is the indicator, and it is the privilege of every engineer to have at least an elementary, if not a thorough knowledge of its principles and working. The time devoted to the study of the indicator, and in its application to the engine, is time well spent, and the end will well repay the student of steam engineering.

Inventor. The indicator was invented and first applied to the steam engine by James Watt, whose restless genius was not satisfied with a mere outside view of his engine as it was running, but he desired to know more about the action of the steam in the cylinder, its pressure at different portions of the stroke, the laws governing its expansion after being out off, etc. Watt's indicator, although crude in its design and construction, contained embodied within it all of the principles of the modern instrument.

Principles. These principles are:

First. The pressure of the steam in the engine cylinder throughout an entire revolution, against a small piston in the cylinder of the indicator, which in turn is controlled or resisted in its movement by a spring of known tension, so as to confine the stroke of the indicator piston within a certain small limit.

Second. The stroke of the indicator piston is communicated by a multiplying mechanism of levers and parallel motion to a pencil moving in a straight line; the distance through which the pencil moves being governed by the pressure in the engine cylinder and the tension of the spring.

Third. By the intervention of a reducing mechanism and a strong cord, the motion of the piston of the

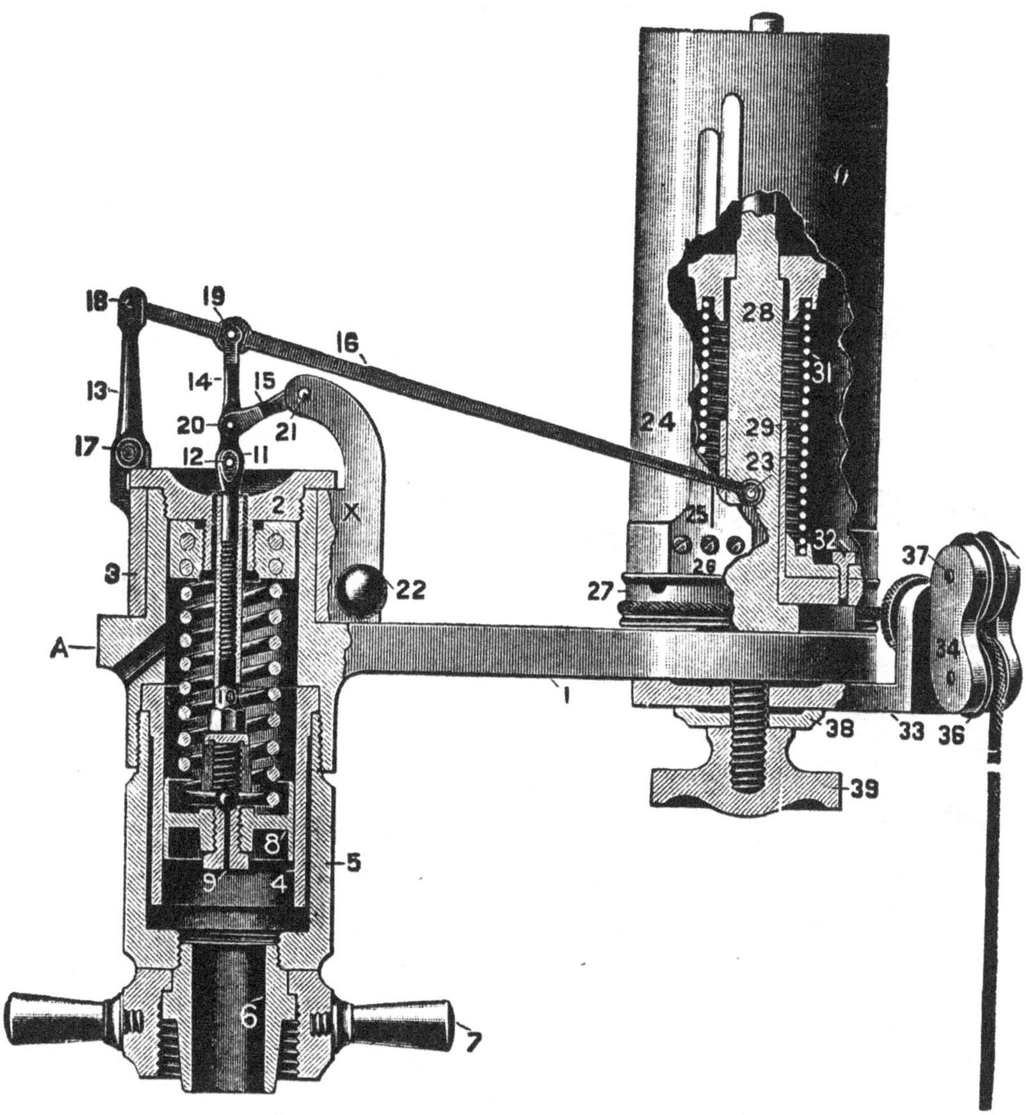

SECTIONAL VIEW CROSBY INDICATOR

engine throughout an entire revolution is communicated to a small drum attached to and forming a part of the indicator. The movement of the drum is rotative and in a direction at right angles to the movement of the pencil. The forward stroke of the engine piston causes the drum to rotate through part of a revolution and at the same time a clock spring connected within the drum is wound up. On the return stroke the motion of the drum is reversed, and the tension of the spring returns the drum to its original position and also keeps the cord taut.

To the outside of the drum a piece of blank paper of suitable size is attached and held in place by two clips. Upon this paper the pencil in its motion up and down traces a complete diagram of the pressures and other interesting events transpiring within the engine cylinder during the revolution of the engine. In fact, the diagram traced upon the paper is the compound result of two concurrent movements. First, that of the pencil, caused by the pressure of the steam against the indicator piston; second, that of the paper drum, caused by, and coincident with, the motion of the engine piston. The upper end of the indicator cylinder is always open to the atmosphere, the steam acting only upon the under side of the small piston, and when the cock connecting the cylinders of the engine and indicator is closed, both ends of the indicator cylinder are open to atmospheric pressure, and the pencil then stands at its neutral position. If now the pencil is held against the paper and the drum rotated either by hand or by connecting it with the cord, a horizontal line will be traced. This line is called the atmospheric line, meaning the line of atmospheric pressure, and it is a very important factor in the study of the diagram.

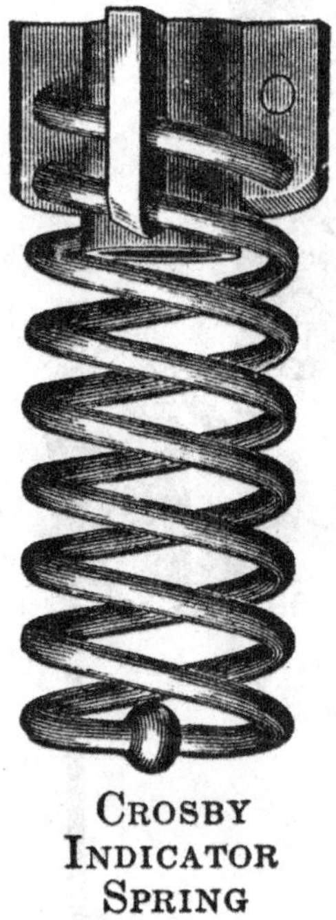

CROSBY INDICATOR SPRING

On a locomotive, the pencil, in tracing the diagram, will not, or at least should not, fall below the atmos-

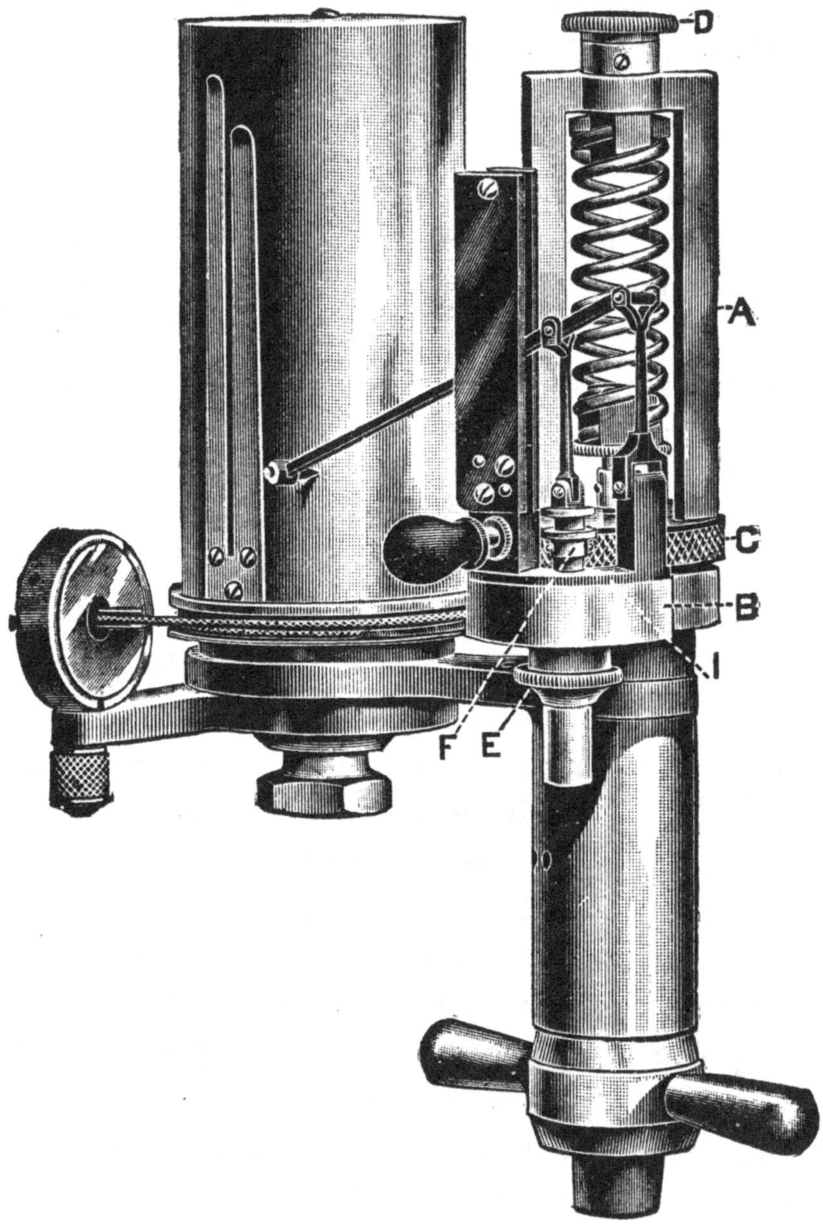

IMPROVED TABOR INDICATOR WITH OUTSIDE CONNECTED SPRING
ASHCROFT MFG. CO., N. Y.

pheric line at any point, but will on the return stroke trace a line called the line of back pressure.

As before stated, the length of stroke of the indicator piston, and the pencil movement as well, is controlled by a spiral steel spring which acts in resistance to the pressure of the steam. These springs are made of different tensions, in order to be suitable to different steam pressures and speeds, and are numbered 20, 40, 60, etc., the number meaning that a pressure per square inch in the engine cylinder corresponding to the number on the spring will cause a vertical movement of the pencil through a distance of one inch. Thus, if a number 20 spring is used and the pressure in the cylinder at the commencement of the stroke is 20 lbs. per square inch, the pencil will be raised one inch, or if the pressure is 30 lbs., the pencil will travel 1½ in., and if there is a vacuum of 20 in. in the condenser, the pencil will drop ½ in. below the atmospheric line, for the reason that 20 in. of vacuum corresponds to a pressure of about 10 lbs. less than atmospheric pressure or an absolute pressure of about 4 lbs. If a 60 spring is used, a pressure of 60 lbs. in the engine cylinder will be required to raise the pencil one inch, or 90 lbs. to raise it 1½ in.

The Ashcroft Manufacturing Co. of New York, makers of the well known Tabor indicator, have recently introduced a new feature in indicator work by connecting the spring on top of the cylinder and in plain view of the operator. This arrangement removes the spring from the influence of direct contact with the steam, and it is subject only to the temperature of the surrounding atmosphere. It is claimed that as a result of this the accuracy of the spring is insured and that no allowance need be made in its manufacture for expansion caused by the high temperature to which it is subject when located within the cylinder. Another good feature of this design is, that the spring can be easily removed without disconnecting any one part of the instrument in case it is desired to change springs. A cut of the improved instrument is herewith presented.

Fig. 99 is a sectional view of the American Thompson improved indicator. Fig. 100 shows the spring. Fig. 101 is a three-way cock for attaching the indicator to the cylinder.

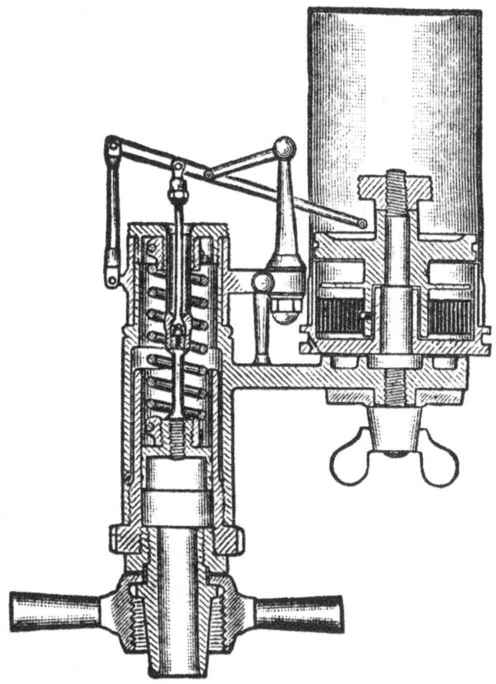

FIGURE 99

Reducing Mechanism. Probably the only practically universal mechanism for reducing the motion of the crosshead is the reducing wheel, a device in which, by the employment of gears and pulleys of different diameters, the motion is reduced to within the compass of the drum, and the device is applicable to almost any make of engine, whether of high or low speed. Some makers of indicators attach the reducing wheel directly to the indicator, thus producing a neat and very convenient arrangement. Fig. 102 shows the indicator complete, with reducing wheel attached.

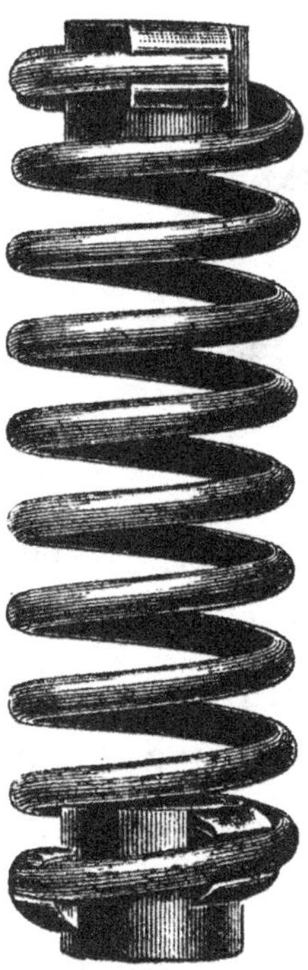

FIGURE 100

Attaching the Indicator. The cylinders of most engines at the present time are drilled and tapped for indicator connections before they leave the shop, which is eminently proper, as no engine builder, or purchaser either, should be satisfied with the performance of a new engine until after it has been accurately tested and adjusted with the indicator.

The main requirements in these connections are that the holes shall not be drilled near the bottom of the cylinder where water is likely to find its way into the pipes, neither should they be in a location where the inrush of steam from the ports will strike them directly, nor where the edge of the piston is liable to

FIGURE 101

partly cover them when at its extreme travel. An engineer before he undertakes to indicate an engine should satisfy himself that all these requirements are fulfilled. Otherwise he is not likely to obtain a true diagram. The cock supplied with the indicator is threaded for one-half inch pipe, and unless the engine has a very long stroke it is the practice to bring the two end connections together at the side or top of the cylinder and at or near the middle of its length, where they can be connected to a three-way cock. The pipe connections should be as short and as free from elbows as possible, in order that the steam may strike the indicator piston as nearly as possible at the same moment that it acts upon the engine piston.

These pipes should always be thoroughly blown out and cleaned, by allowing the steam to blow through

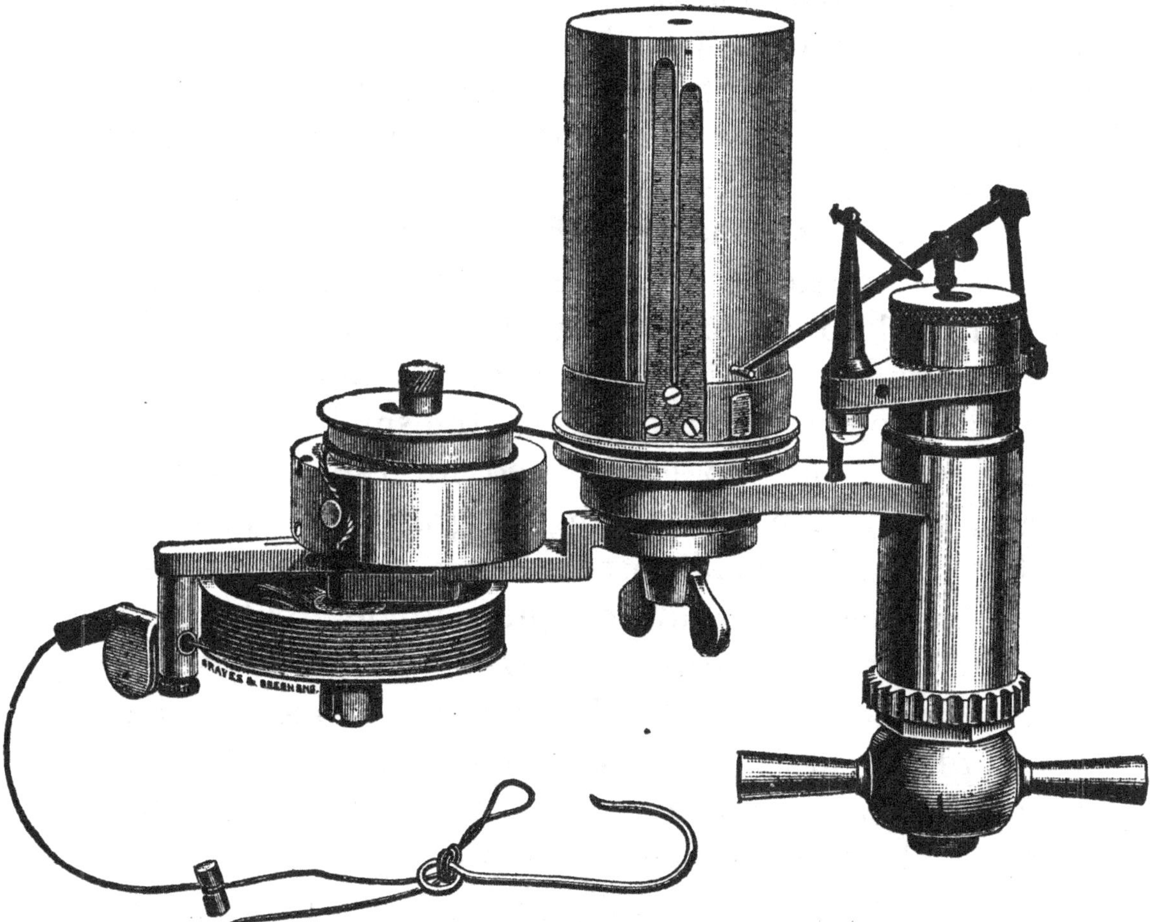

FIGURE 102

the open three-way cock during several revolutions of the engine, before connecting the indicator. If this is not done there is a moral certainty that dirt and grit will get into the cylinder of the indicator and cause it to work badly, and give diagrams that are misleading. As before stated, the height of the diagram depends upon the tension or number of the spring. It is a convenient practice to select a spring numbered one-half of the boiler pressure, as, for instance, suppose gauge pressure or boiler pressure is 200 lbs. per sq. in., then a 100 spring would give a diagram 2 inches in height, which is a convenient height. As to the length of the diagram, this is regulated by adjustment of the cord in its travel, by means of the reducing wheel. Any length of diagram up to four inches may be obtained, but two and a half to three inches is a very good length for analysis.

Care of the Instrument. The indicator should be cleaned and oiled both before and after using. The best material for wiping it is a clean piece of old soft muslin of fine texture, as there is not so much liability of lint sticking to, or getting into, the small joints.

Good clock oil should be used for the joints and springs, and just before taking diagrams it is a good practice to rub a small portion of cylinder oil on the piston and on the inside of the cylinder, but when about to put the instrument away, these should be cleaned and oiled with clock oil also. None but the best cord should be used for connecting the reducing wheel with the crosshead, as a cord that is liable to stretch will cause trouble. Suitable cord, and also blank diagrams, can generally be obtained from firms engaged in manufacturing and selling indicators. After the indicator has been screwed on to the cock connecting with the pipe, the cord must be adjusted to the proper length before hooking it on to the drum. This must be done while the engine is running, by taking hold of the loop on the cord connected with the crosshead with one hand, and with the other hand grasp the hook on the cord attached to the reducing wheel; then, by holding the two ends near each other during a revolution or two of the crank pin, it will be seen whether the long cord needs to be lengthened or shortened. Care should be exercised in placing the paper on the drum to see that it is stretched tight and firmly held by the clips. The pencil point, having been first sharpened by rubbing it on a piece of fine emery cloth or sandpaper, should be adjusted by means of the pencil stop with which all indicators should be provided, so that it will have just sufficient bearing against the paper to make a fine, plain mark. If the pencil bears too hard on the paper it will cause unnecessary friction and the diagram will be distorted. The best method of ascertaining this fact and also whether the travel of the drum is equally divided between the stops, is to place a blank diagram on the drum, connect the cord and while the engine makes a revolution hold the pencil against the paper. Then unhook the cord, remove the paper and if the travel of the drum is not divided correctly it can be changed.

Having thus arranged all the preliminary details, place a fresh blank on the drum, being careful to keep the pencil out of contact with it, connect the cord, open the cock admitting steam to the indicator, and after the pencil has made a few strokes to allow the cylinder to become warmed up, then gently swing it around to the paper drum and hold it there while the engine makes a complete revolution. Then move the pencil clear of the paper, close the cock and unhook the cord. Now trace the atmospheric line by holding the pencil against the paper while the drum is revolved by hand. This method of tracing the atmospheric line is preferable to that of tracing it immediately after closing the cock and while the drum is still being moved by the engine, for the reason that there is not so much liability of getting the atmospheric line too high owing to the presence of a slight pressure of steam remaining under the indicator piston for a second or two just after closing the cock; also the line drawn by hand will be longer than one drawn while the drum is moved by the motion of the engine and will therefore be more readily distinguished from the line of back pressure.

Having secured a truthful diagram, it now remains to take as many as are desired, and they should follow each other as rapidly as possible, in order that each pair of diagrams may be taken under the same conditions of initial pressure, cut-off, etc. In order to get accurate results, the operator should have an assistant posted in the cab, whose duty will be to watch the steam gauge, and see that other conditions are the same at least during the time a pair of cards is being taken. As soon as the diagrams are taken, the following data should be noted upon them: the end of cylinder, whether head end, or crank end, boiler pressure, revolutions per minute, miles per hour, throttle opening, cut-off. Other data, such as mean effective pressure, back pressure, indicated horse-power, and steam per indicated horse-power per hour, may be ascertained by an analysis of the diagrams, and should also be noted upon the back of each pair of diagrams after they have been found by calculation. The diagrams should be numbered, also, as they are taken.

The taking of indicator diagrams from locomotives has of late years been greatly facilitated by the use of electrical apparatus whereby any number of diagrams may be taken simultaneously. This is certainly a great improvement over the old method of hand manipulation, especially for high speed engines.

In order to facilitate the study and analysis of indicator diagrams, the following definitions of technical terms, some of which have already been explained in another part of this book, are here given.

Absolute Pressure. Pressure reckoned from a perfect vacuum. It equals the boiler pressure plus the atmospheric pressure.

Boiler Pressure or Gauge Pressure. Pressure above the atmospheric pressure as shown by the steam gauge.

Initial Pressure. Pressure in the cylinder at the beginning of the stroke.

Terminal Pressure (T. P). The pressure that would exist in the cylinder at the end of the stroke provided the exhaust valve did not open until the stroke was entirely completed. It may be graphically illustrated on the diagram by extending the expansion curve by hand to the end of the stroke. It is found theoretically by dividing the pressure at point of cut-off by the ratio of expansion. Thus, absolute pressure at cut-off = 100 lbs., ratio of expansion = 5; then 100 ÷ 5 = 20 lbs., absolute terminal pressure.

Mean Effective Pressure (M. E. P.). The average pressure acting upon the piston throughout the stroke minus the back pressure.

Back Pressure. Pressure which tends to retard the forward stroke of the piston. Indicated on the diagram from a non-condensing engine by the height of the back pressure line above the atmospheric line. In a condensing engine the degree of back pressure is shown by the height of the back pressure line above an imaginary line representing the pressure in the condenser corresponding to the degree of vacuum in inches, as shown by the vacuum gauge.

Total or Absolute Back Pressure, in either a condensing or non-condensing engine, is that indicated on the diagram by the height of the line of back pressure above the line of perfect vacuum.

Ratio of Expansion. The proportion that the volume of steam in the cylinder at point of release bears to the volume at cut-off. Thus, if the point of cut-off is at one-fifth of the stroke, and release does not take place until the end of the stroke, the ratio of expansion, or in other words, the number of expansions, is 5. When the T. P. is known the ratio of expansion may be found by dividing the initial pressure by the T. P.

Wire-Drawing. When through insufficiency of valve opening, or contracted ports, the steam is prevented from following up the piston at full initial pressure until the point of cut-off is reached, it is said to be wire-drawn. It is indicated on the diagram by a gradual inclination downwards of the steam line from the admission line to the point of cut-off. Too small a steam pipe from boiler to engine will also cause wire-drawing and fall of pressure.

Condenser Pressure may be defined as the pressure existing in the condenser of an engine, caused by the lack of a perfect vacuum; as, for instance, with a vacuum of 25 in. there will still remain the pressure due to the 5 in. which is lacking. This will be about 2.5 lbs.

Vacuum. That condition existing within a closed vessel from which all matter, including air, has been expelled. It is measured by inches in a column of mercury contained within a glass tube a little over 30 in. in height, having its lower end open and immersed in a small open vessel filled with mercury. The upper end of the glass tube is connected with the vessel in which the vacuum is to be produced. When no vacuum exists the mercury will leave the tube and fill the lower vessel. When a vacuum is maintained in the condenser, or other vessel, the mercury will rise in the glass tube to a height corresponding to the degree of vacuum. If the mercury rises to the height of 30 in. it indicates a perfect vacuum, which means the absence of all pressure within the vessel, but this condition is never realized in practice; the nearest approach to it being about 28 in.

For purposes of convenience the mercurial vacuum gauge is not generally used, it having been replaced by the Bourdon spring gauge, although the mercury gauge is used for testing.

The vacuum in a condenser is generally maintained by an air pump, although it can be produced and maintained by the mere condensation of the steam as it enters the condenser by allowing a spray of cold water to strike it. The steam when it first enters the condenser drives out the air and the vessel is filled with steam, which, when condensed, occupies about 1,600 times less space than it did before being condensed; hence a partial vacuum is produced.

While the vacuum in a condenser cannot be consid ered as power at all, yet it occupies the anomalous position of increasing, by its presence, the capacity of the engine for doing work. This is owing to the fact that the atmospheric pressure or resistance which is always ahead of the piston in a non-condensing engine is, in the case of a condensing engine, removed to a degree corresponding to the height of the vacuum, thus making available just so much more of the pressure behind the piston. Thus, if the average steam pressure throughout the stroke is 30 lbs. and there is a vacuum of 26 in. maintained in the condenser, there will be 13 lbs. of resistance per square inch removed from in front of the piston, thus making available 30 + 13 = 43 lbs. pressure per square inch.

Absolute Zero has been fixed by calculation at 461.2° below the zero of the Fahrenheit scale.

Piston Displacement. The space or volume swept through by the piston in a single stroke. Found by multiplying the area of piston by length of stroke.

Piston Clearance. The distance between the piston and cylinder head when the piston is at the end of the stroke.

Steam Clearance, Ordinarily Termed Clearance. The space between the piston at the end of the stroke and the valve face. It is reckoned in per cent of the total piston displacement.

Horse-Power (H. P.). 33,000 pounds raised one foot high in one minute of time.

Indicated Horse-Power (I. H. P.). The horse-power as shown by the indicator diagram. It is found as follows:

Area of piston in square inches × M. E. P. × piston speed in feet ÷ 33,000.

Piston Speed. The distance in feet traveled by the piston in one minute. It is the product of twice the length of stroke expressed in feet multiplied by the number of revolutions per minute.

R. P. M. Revolutions per minute.

Net Horse-Power. I. H. P. minus the friction of the engine.

Compression. The action of the piston as it nears

the end of the stroke, in reducing the volume and raising the pressure of the steam retained in the cylinder ahead of the piston by the closing of the exhaust valve.

Boyle's or Mariotte's Law of Expanding Gases. "The pressure of a gas at a constant temperature varies inversely as the space it occupies." Thus, if a given volume of gas is confined at a pressure of 50 lbs. per square inch and it is allowed to expand to twice its volume, the pressure will fall to 25 lbs. per square inch.

Adiabatic Curve. A curve representing the expansion of a gas which loses no heat while expanding. Sometimes called the curve of no transmission.

Isothermal Curve. A curve representing the expansion of a gas having a constant temperature but partially influenced by moisture, causing a variation in pressure according to the degree of moisture or saturation. It is also called the theoretical expansion curve.

Expansion Curve. The curve traced upon the diagram by the indicator pencil, showing the actual expansion of the steam in the cylinder.

First Law of Thermodynamics. Heat and mechanical energy are mutually convertible.

Power. The rate of doing work, or the number of foot-pounds exerted in a given time.

Unit of Work. The foot-pound, or the raising of one pound weight one foot high.

First Law of Motion. All bodies continue either in a state of rest or of uniform motion in a straight line, except in so far as they may be compelled by impressed forces to change that state.

Work. Mechanical force or pressure cannot be considered as work unless it is exerted upon a body and causes that body to move through space. The product of the pressure multiplied by the distance passed through and the time thus occupied is work.

Momentum. Force possessed by bodies in motion, or the product of mass and density.

Dynamics. The science of moving powers or of matter in motion, or of the motion of bodies that mutually act upon each other.

Force. That which alters the motion of a body, or puts in motion a body that was at rest.

Maximum Theoretical Duty of Steam is the product of the mechanical equivalent of heat, viz., 778 ft. lbs., multiplied by the total heat units in a pound of steam. Thus, in one pound of steam at 212° reckoned from 32° the total heat equals 1,146.6 heat units. Then $778 \times 1{,}146.6$ equals 892,054.8 ft. lbs. = maximum duty.

Steam Efficiency may be expressed as follows:

$$\frac{\text{Heat converted into useful work}}{\text{Heat expended}}$$ and maximum efficiency can only be attained by using steam at as high an initial pressure as is consistent with safety and at as large a ratio of expansion as possible. The percentage of efficiency of steam used at atmospheric pressure in a non-expansive engine is very low; as, for instance, the heat expended in the evaporation of one pound of water at 32° into steam at atmospheric pressure = 1,146.6 heat units, and the volume of steam so generated = 26.36 cu. ft.

One cubic foot of steam at 212° contains energy equal to $144 \times 14.7 = 2{,}116.8$ ft. lbs., and 26.36 cu. ft. = $2{,}116.8 \times 26.36 = 55{,}798.84$ ft. lbs., which divided by the mechanical equivalent of heat, viz., 778 ft. lbs. = 71.72 heat units, available for useful work. The per cent of efficiency, therefore, is $\frac{71.72 \times 100}{1{,}146.6} = 6.28$ per cent. But suppose the initial pressure to have been 200 lbs. absolute, and that the steam is allowed to expand to thirty times its original volume. The heat expended in evaporating a pound of water at 32° into steam at 200 lbs. absolute pressure = 1,198.3 heat units, and the volume of steam so generated = 2.27 cu. ft. The average pressure during expansion would be 29.34 lbs. per square inch and the volume when expanded thirty times would equal 2.27 × 30 = 68.1 cu. ft.

One cubic foot of steam at 29.34 lbs. pressure equals 144 × 29.34 = 4,224.96 ft. lbs., and 68.1 cu. ft. will equal 4,224.96 × 68.1 = 287,719.7 ft. lbs. of energy, which divided by the equivalent, 778, equals 370.2 heat units, and the per cent of efficiency will be $\frac{370.2 \times 100}{1,198.3} = 30.8$ per cent.

Engine Efficiency. If the engine is considered merely as a machine for converting into useful work the heat energy in the steam regardless of the cost of fuel, its efficiency may be expressed as follows:

$$\frac{\text{Heat converted into useful work}}{\text{Total heat received in the steam}}$$

Example. Assume an engine to be receiving steam at 95 lbs. absolute pressure, that the consumption of dry steam per horse-power per hour equals 20 lbs., that the friction of the engine amounts to 15 per cent, and that the temperature of the feed water is raised from 60° to 170° by utilizing a portion of the exhaust.

In a pound of steam at 95 lbs. absolute there are 1,180.7 heat units, and in a pound of water at 170° there are 138.6 units of heat, but 28.01 of these heat units were in the water at its initial temperature of 60°. Therefore the total heat added to the water by the exhaust steam equals 138.6 − 28.01 = 110.59 heat units, and the total heat in each pound of steam to be charged up to the engine is 1,180.7 − 110.59 = 1,070.11, and the total for each horse-power developed per hour will be 1,070.11 × 20 = 21,402.2 heat units.

A horse-power equals 33,000 ft. lbs. per minute, or sixty times 33,000 = 1,980,000 ft. lbs. per hour. From this must be deducted 15 per cent for friction of the engine, leaving 1,683,000 ft. lbs. for useful work. Dividing this by the equivalent, viz., 778 ft. lbs., gives 2,163.2 heat units as the heat converted into one horse-power of work in one hour, and the percentage of efficiency of the engine will be $\frac{2,163.2 \times 100}{21,402.2} = 10.1$ per cent.

Efficiency of the Plant as a Whole includes boiler and engine efficiency and is to be figured upon the basis of

$$\frac{\text{Heat converted into useful work}}{\text{Calorific or heat value of fuel}}$$

Horse-Power Constant of an engine is found by multiplying the area of the piston in square inches by the speed of the piston in feet per minute and dividing the product by 33,000. It is the power the engine would develop with one pound mean effective pressure. To find the horse-power of the engine, multiply the M. E. P. of the diagram by this constant.

Logarithms. A series of numbers having a certain relation to the series of natural numbers, by means of which many arithmetical operations are made comparatively easy. The nature of the relation will be understood by considering two simple series, such as the following, one proceeding from unity in geometrical progression and the other from 0 in arithmetical progression

Geom. series, 1, 2, 4, 8, 16, 32, 64, 128, 256, 512, etc.

Arith. series, 0, 1, 2, 3, 4, 5, 6, 7, 8, 9, etc.

Here the ratio of the geometrical series is 2 and any term in the arithmetical series expresses how often 2 has been multiplied into 1 to produce the corresponding term of the geometrical series. Thus, in proceeding from 1 to 32 there have been 5 steps or multiplications by the ratio 2; in other words, the ratio of 32 to 1 is compounded 5 times of the ratio of 2 to 1. The above is the basic principle upon which common logarithms are computed.

Hyperbolic Logarithms. Used in figuring the M. E. P. of a diagram from the ratio of expansion and the initial pressure. Thus, hyperbolic logarithm of ratio of expansion + 1 multiplied by absolute initial pressure and divided by ratio of expansion = mean forward pressure. From this deduct total back pressure and the remainder will be mean effective pressure. The hyperbolic logarithm is found by multiplying the common logarithm by the constant 2.302585. Table 14 gives the hyperbolic logarithms of numbers usually required in calculations of the above nature.

Steam Consumption per Horse-Power per Hour. The weight in pounds of steam exhausted into the atmosphere or into the condenser in one hour divided by the horse-power developed. It is determined from the diagram by selecting a point in the expansion curve just previous to the opening of the exhaust valve and measuring the absolute pressure at that point. Then the piston displacement up to the point selected, plus the clearance space, expressed in cubic feet, will give the volume of steam in the cylinder, which multiplied by the weight per cubic foot of steam at the pressure as measured will give the weight of steam consumed during one stroke. From this should be deducted the

TABLE 14

HYPERBOLIC LOGARITHMS.

No.	Log.	No.	Log.	No.	Log.	No.	Log.	No.	Log.
1.01	0.0099	3.00	1.0986	5.00	1.6094	7.00	1.9459	9.00	2.1972
1.05	0.0487	3.05	1.1151	5.05	1.6194	7.05	1.9530	9.05	2.2028
1.10	0.0953	3.10	1.1341	5.10	1.6292	7.10	1.9600	9.10	2.2083
1.15	0.1397	3.15	1.1474	5.15	1.6390	7.15	1.9671	9.15	2.2137
1.20	0.1823	3.20	1.1631	5.20	1.6486	7.20	1.9740	9.20	2.2192
1.25	0.2231	3.25	1.1786	5.25	1.6582	7.25	1.9810	9.25	2.2246
1.30	0.2623	3.30	1.1939	5.30	1.6677	7.30	1.9879	9.30	2.2310
1.35	0.3001	3.35	1.2090	5.35	1.6771	7.35	1.9947	9.35	2.2354
1.40	0.3364	3.40	1.2238	5.40	1.6864	7.40	2.0015	9.40	2.2407
1.45	0.3715	3.45	1.2384	5.45	1.6956	7.45	2.0018	9.45	2.2460
1.50	0.4054	3.50	1.2527	5.50	1.7047	7.50	2.0149	9.50	2.2513
1.55	0.4382	3.55	1.2669	5.55	1.7138	7.55	2.0215	9.55	2.2565
1.60	0.4700	3.60	1.2809	5.60	1.7228	7.60	2.0281	9.60	2.2618
1.65	0.5007	3.65	1.2947	5.65	1.7316	7.65	2.0347	9.65	2.2670
1.70	0.5306	3.70	1.3083	5.70	1.7405	7.70	2.0412	9.70	2.2721
1.75	0.5596	3.75	1.3217	5.75	1.7491	7.75	2.0477	9.75	2.2773
1.80	0.5877	3.80	1.3350	5.80	1.7578	7.80	2.0541	9.80	2.2824
1.85	0.6151	3.85	1.3480	5.85	1.7664	7.85	2.0605	9.85	2.2875
1.90	0.6418	3.90	1.3610	5.90	1.7750	7.90	2.0668	9.90	2.2925
1.95	0.6678	3.95	1.3737	5.95	1.7834	7.95	2.0731	9.95	2.2976
2.00	0.6931	4.00	1.3863	6.00	1.7918	8.00	2.0794	10.00	2.3026
2.05	0.7178	4.05	1.3987	6.05	1.8000	8.05	2.0857	10.25	2.3273
2.10	0.7419	4.10	1.4010	6.10	1.8083	8.10	2.0918	10.50	2.3514
2.15	0.7654	4.15	1.4231	6.15	1.8164	8.15	2.0988	10.75	2.3749
2.20	0.7885	4.20	1.4351	6.20	1.8245	8.20	2.1041	11.00	2.3979
2.25	0.8110	4.25	1.4469	6.25	1.8326	8.25	2.1102	12.00	2.4849
2.30	0.8329	4.30	1.4586	6.30	1.8405	8.30	2.1162	13.00	2.5626
2.35	0.8544	4.35	1.4701	6.35	1.8484	8.35	2.1222	14.00	2.6390
2.40	0.8755	4.40	1.4816	6.40	1.8563	8.40	2.1282	15.00	2.7103
2.45	0.8961	4.45	1.4929	6.45	1.8640	8.45	2.1342	16.00	2.7751
2.50	0.9163	4.50	1.5040	6.50	1.8718	8.50	2.1400	17.00	2.8332
2.55	0.9361	4.55	1.5151	6.55	1.8795	8.55	2.1459	18.00	2.8903
2.60	0.9555	4.60	1.5260	6.60	1.8870	8.60	2.1518	19.00	2.9444
2.65	0.9746	4.65	1.5369	6.65	1.8946	8.65	2.1576	20.00	2.9957
2.70	0.9932	4.70	1.5475	6.70	1.9021	8.70	2.1633	21.00	3.0445
2.75	1.0116	4.75	1.5581	6.75	1.9095	8.75	2.1690	22.00	3.0910
2.80	1.0296	4.80	1.5686	6.80	1.9169	8.80	2.1747	23.00	3.0355
2.85	1.0473	4.85	1.5790	6.85	1.9242	8.85	2.1804	24.00	3.1780
2.90	1.0647	4.90	1.5892	6.90	1.9315	8.90	2.1860	25.00	3.2189
2.95	1.0818	4.95	1.5994	6.95	1.9387	8.95	2.1916	30.00	3.3782

steam saved by compression as shown by the diagram, in order to get a true measure of the economy of the engine. Having thus determined the weight of steam consumed for one stroke, multiply it by twice the number of strokes per minute and by 60, which will give the total weight consumed per hour. This divided by the horse-power will give the rate per horse-power per hour.

Cylinder Condensation and Reëvaporation. When the exhaust valve opens to permit the exit of the steam there is a perceptible cooling of the walls of the cylinder, especially in condensing engines when a high vacuum is maintained. This results in more or less condensation of the live steam admitted by the opening of the steam valve; but if the exhaust valve is caused to close at the proper time so as to retain a portion of the steam to be compressed by the piston on the return stroke, a considerable portion of the water caused by condensation will be reëvaporated into steam by the heat and consequent rise in pressure caused by compression.

Ordinates. Parallel lines drawn at equal distances apart across the face of the diagram, and perpendicular to the atmospheric line. They serve as a guide to facilitate the measurement of the average forward pressure throughout the stroke, or the pressure at any point of the stroke if desired.

Eccentric. A mechanical device used in place of a crank for converting rotary into reciprocating motion. An eccentric is in fact a form of crank in which the crank pin, corresponding to the eccentric sheave, embraces the shaft, but owing to the great leverage at which the friction between the sheave and the strap acts, compared with its short turning leverage, it can only be used to advantage for the purpose named above.

Eccentric Throw is the distance from the center of the eccentric to the center of the shaft. This definition also applies to the term "radius of eccentricity."

Eccentric Position. The location of the highest poin: of the eccentric relative to the center of the crank pin, measured or expressed in degrees.

Angular Advance. The distance that the high point of the eccentric is set ahead of a line at right angles with the crank. In other words, the lap angle plus the lead angle. If a valve had neither lap nor lead, the position of the high point of the eccentric would be on a line at right angles with the crank; as, for instance, the crank being at 0° the eccentric would stand at 90°.

Valve Travel. The distance covered by the valve in its movement. It equals twice the throw of the eccentric.

Lap. The amount that the ends of the valve project over the edges of the ports when the valve is at mid-travel.

Outside or Steam Lap. The amount that the end of the valve overlaps or projects over the outside edge of the steam port.

Inside Lap. The lap of the inside or exhaust edge of the valve over the inside edge of the port.

Lead. The amount that the port is open when the crank is on the dead center. The object of giving a valve lead is to supply a cushion of live steam which, in conjunction with that already confined in the clearance space by compression, shall serve to bring the moving parts of the engine to rest quietly at the end of the stroke, and also quicken the action of the piston in beginning the return stroke.

Compression. Closing of the exhaust passage before the steam is entirely exhausted from the cylinder. A certain quantity of steam is thus compressed into the clearance space.

Table 15, giving areas and circumferences of circles, is here inserted, for the reason that in the study of indicator diagrams there is very often occasion for reference to such a table.

In the following analysis of indicator diagrams all

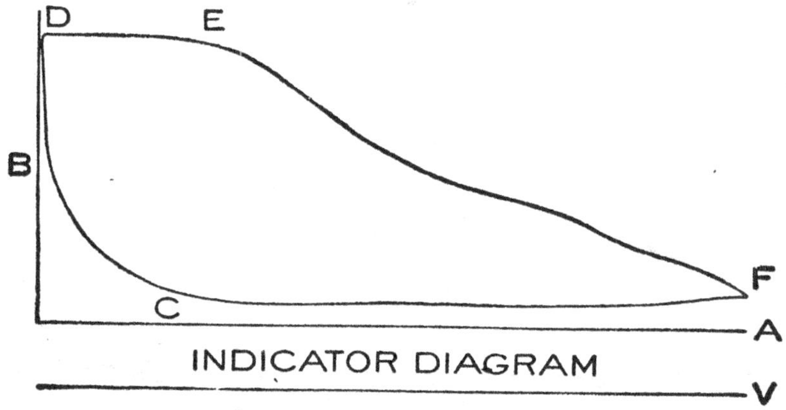

FIGURE 103

of the illustrations are reproductions of actual diagrams taken under ordinary working conditions.

Fig. 103 shows a sample diagram taken from a locomotive running at a speed of 29 miles per hour, and cutting off at 5½ inches. By reference to the letters, the different lines and points into which an indicator diagram is divided may be readily distinguished. The line V indicates the base line, or line of perfect vacuum, from which pressures are measured, especially in calculations of steam consumed per horse-power hour. This line is drawn at a point 14.7 lbs. below the atmospheric line A, as indicated by measurements made with the scale adapted to the spring used in taking the diagram. The method of drawing the line of atmospheric pressure has already been described. It is from this line that the mean effective pressure of the steam upon the piston during the stroke is estimated in all calculations of diagrams taken from locomotives and other non-condensing engines.

Admission at the beginning of the stroke is shown at B, and from B to D is the admission line. From D to E is the steam line. E is the point of cut-off, and from E to F is the expansion curve. F is the point of release or exhaust opening, and from this point to C

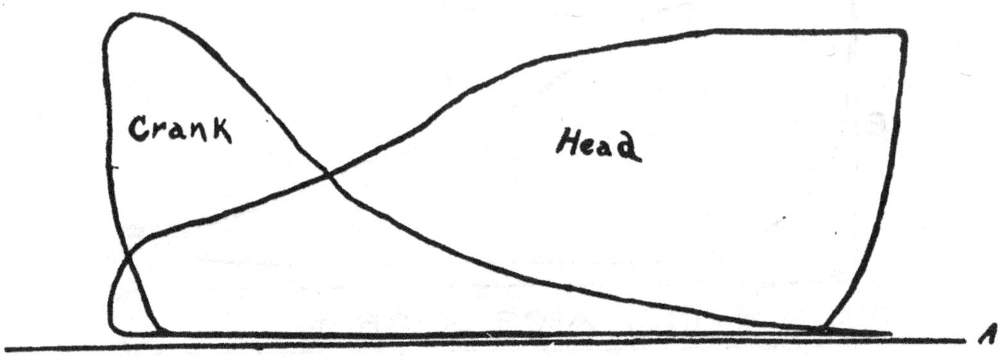

FIGURE 104

is the line of back pressure or counter pressure, and the amount of this pressure depends upon the height of this line above line A. Compression begins at C, and from this point to B is the compression curve.

Fig. 104 shows bad valve adjustment, the engine doing by far the largest portion of the work in the head end of the cylinder.

In order to illustrate the process of ascertaining the M. E. P. without dividing the diagram into ordinates, the following computation is given, together with rules, etc. In this process two important factors are necessary, viz., the absolute initial pressure and the absolute

Table 15

Areas and Circumferences of Circles.

Diam.	Area.	Circum.	Diam.	Area.	Circum.	Diam.	Area.	Circum.
.25	.049	.7854	15.5	188.692	48.694	31	754.769	97.389
.5	.1963	1.5708	16	201.062	50.265	31.25	766.992	98.175
1.0	.7854	3.1416	16.25	207.394	51.051	31.5	799.313	98.968
1.25	1.2271	3.9270	16.5	213.825	51.836	32	804.249	100.53
1.5	1.7671	4.7124	17	226.980	53.407	32.25	816.86	101.31
2	3.1416	6.2832	17.25	233.705	54.192	33	855.30	103.67
2.25	3.9760	7.0686	17.5	240.520	54.978	33.25	868.30	104.45
2.5	4.9087	7.8540	18	254.469	56.548	33.5	881.41	105.24
3	7.0686	9.4248	18.25	261.587	57.334	34	907.92	106.81
3.25	8.2957	10.210	18.5	268.803	58.119	34.25	921.32	107.60
3.5	9.6211	10.995	19	283.529	59.690	34.5	934.82	108.38
4	12.566	12.566	19.25	291.039	60.475	35	962.11	106.95
4.25	14.186	13.351	19.5	298.648	61.261	35.25	975.90	110.74
4.5	15.904	14.137	20	314.160	62.832	35.5	989.80	111.52
5	19.635	15.708	20.25	322.063	63.617	36	1017.8	113.09
5.25	21.647	16.493	20.5	330.064	64.402	36.25	1032.06	113.88
5.5	23.758	17.278	21	346.361	65.973	36.5	1046.35	114.66
6	28.274	18.849	21.25	354.657	66.759	37	1075.21	116.23
6.25	30.679	19.635	21.5	363.051	67.544	37.25	1089.79	117.01
6.5	33.183	20.420	22	380.133	69.115	37.5	1104.46	117.81
7	38.484	21.991	22.25	388.822	69.900	38	1134.11	119.38
7.25	41.282	22.776	22.5	397.608	70.686	38.25	1149.08	120.16
7.5	44.178	23.562	23	415.476	72.256	38.5	1164.15	120.95
8	50.265	25.132	23.25	424.557	73.042	39	1194.59	122.52
8.25	53.456	25.918	23.5	433.731	73.827	39.25	1209.95	123.30
8.5	56.745	26.703	24	452.390	75.398	39.5	1225.42	124.09
9	63.617	28.274	24.25	461.864	76.183	40	1256.64	125.66
9.25	67.200	29.059	24.5	471.436	76.969	40.25	1272.39	126.44
9.5	70.882	29.845	25	490.875	78.540	40.5	1288.25	127.23
10	78.540	31.416	25.25	500.741	79.325	41	1320.25	128.80
10.25	82.516	32.201	25.5	510.706	80.110	41.25	1336.40	129.59
10.5	86.590	32.986	26	530.930	81.681	41.5	1352.65	130.37
11	95.033	34.557	26.25	541.189	82.467	42	1385.44	131.94
11.25	99.402	35.343	26.5	551.547	83.252	42.25	1401.98	132.73
11.5	103.869	36.128	27	572.556	84.823	42.5	1418.62	133.51
12	113.097	37.699	27.25	583.208	85.608	43	1452.20	135.08
12.25	117.859	38.484	27.5	593.958	86.394	43.25	1469.13	135.87
12.5	122.718	39.270	28	615.753	87.964	43.5	1486.17	136.65
13	132.732	40.840	28.25	626.798	88.750	44	1520 53	138.23
13.25	137.886	41.620	28.5	637.941	89.535	44.25	1537.86	139.01
13.5	143.130	42.411	29	660.521	91.106	44.5	1555.28	139.80
14	153.938	43.982	29.25	671.958	91.891	45	1590.43	141.37
14.25	159.485	44.767	29.5	683.494	92.677	45.25	1608.15	142.15
14.5	165.130	45.553	30	706.860	94.248	45.5	1625.97	142.94
15	176.715	47.124	30.25	718.690	95.033	46	1661.90	144.51
15.25	182.654	47.909	30.5	730.618	95.818	46.25	1680.01	145.29

TABLE 15—*Continued*

Diam.	Area.	Circum.	Diam.	Area.	Circum.	Diam.	Area.	Circum.
46.5	1698.23	146.08	62.25	3043.47	195.56	78	4778.37	245.04
47	1734.94	147.65	62.5	3067.96	196.35	78.25	4809.05	245.83
47.25	1753.45	148.44	63	3117.25	197.92	78.5	4839.83	246.61
47.5	1772.05	149.22	63.25	3142.04	198.71	79	4901.68	248.19
48	1809.56	150.79	63.5	3166.92	199.50	79.25	4932.75	248.97
48.25	1828.46	151.58	64	3216.99	201.06	79.5	4963.92	249.76
48.5	1847.45	152.36	64.25	3242.17	201.85	80	5026.56	251.33
49	1885.74	153.93	64.5	3267.46	202.68	80.5	5089.58	252.90
49.25	1905.03	154.72	65	3318.31	204.20	81	5153.00	254.47
49.5	1924.42	155.50	65.25	3343.88	204.99	81.5	5216.82	256.04
50	1963.50	157.08	65.5	3369.56	205.77	82	5281.02	257.61
50.25	1983.18	157.86	66	3421.20	207.34	82.5	5345.62	259.18
50.5	2002.96	158.65	66.25	3447.16	208.13	83	5410.62	260.75
51	2042.82	160.22	66.5	3473.23	208.91	83.5	5476.00	262.32
51.25	2062.90	161.00	67	3525.66	210.49	84	5541.78	263.89
51.5	2083.07	161.79	67.25	3552.01	211.27	84.5	5607.95	265.46
52	2123.72	163.36	67.5	3573.47	212.06	85	5674.51	267.04
52.25	2144.19	164.14	68	3631.68	213.63	85.5	5741.47	268.60
52.5	2164.75	164.19	68.25	3658.44	214.41	86	5808.81	270.17
53	2206.18	166.50	68.5	3685.29	215.20	86.5	5876.55	271.75
53.25	2227.05	167.29	69	3739.28	216.77	87	5944.66	273.32
53.5	2248.01	168.07	69.25	3766.43	217.55	87.5	6013.21	274.89
54	2290.22	169.64	69.5	3793.67	218.34	88	6082.13	276.46
54.25	2311.48	170.43	70	3848.46	219.91	88.5	6151.44	278.03
54.5	2332.83	171.21	70.25	3875.99	220.70	89	6221.15	279.60
55	2375.83	172.78	70.5	3903.63	221.48	89.5	6291.25	281.17
55.25	2397.48	173.57	71	3959.20	223.05	90	6371.64	282.74
55.5	2419.22	174.35	71.25	3987.13	223.84	90.5	6432.62	284.31
56	2463.01	175.92	71.5	4015.16	224.62	91	6503.89	285.88
56.25	2485.05	176.71	72	4071.51	226.19	91.5	6573.56	287.46
56.5	2507.19	177.5	72.25	4099.83	226.98	92	6647.62	289.03
57	2551.76	179.07	72.5	4128.25	227.75	92.5	6720.07	290.60
57.25	2574.19	179.85	73	4185.39	229.34	93	6792.92	292.17
57.5	2596.72	180.64	73.25	4214.11	230.12	93.5	6866.16	293.74
58	2642.08	182.21	73.5	4242.92	230.91	94	6939.79	295.31
58.25	2664.91	182.99	74	4300.85	232.48	94.5	7013.81	296.88
58.5	2687.83	183.78	74.25	4329.95	233.26	95	7088.23	298.45
59	2733.97	185.35	74.5	4359.16	234.05	95.5	7163.04	300.02
59.25	2757.19	186.14	75	4417.87	235.62	96	7238.25	301.59
59.5	2780.51	186.92	75.25	4447.37	236.40	96.5	7313.80	303.16
60	2827.44	188.49	75.5	4476.97	237.19	97	7389.81	304.73
60.25	2851.05	189.28	76	4536.37	238.76	97.5	7466.22	306.30
60.5	2874.76	190.06	76.25	4566.36	239.55	98	7542.89	307.88
61	2922.47	191.64	76.5	4596.35	240.33	98.5	7620.09	309.44
61.25	2946.47	192.42	77	4656.63	241.90	99	7697.70	311.02
61.5	2970.57	193.21	77.25	4686.92	242.69	99.5	7775.63	312.58
62	3019.07	194.78	77.5	4717.30	243.47	100	7854.00	314.16

terminal pressure, and they can both be obtained from the diagram by measuring with the scale adapted to the spring used. Thus, in Fig. 105 the absolute initial pressure measured from the line of perfect vacuum V to line B is 77 lbs., and the absolute terminal pressure measured from V to line B′ is 21 lbs. The ratio, or number of expansions, is found thus:

Rule. Divide the absolute initial pressure by the absolute terminal pressure; thus, $77 \div 21 = 3.65$ = number of expansions.

Second. Find mean forward pressure.

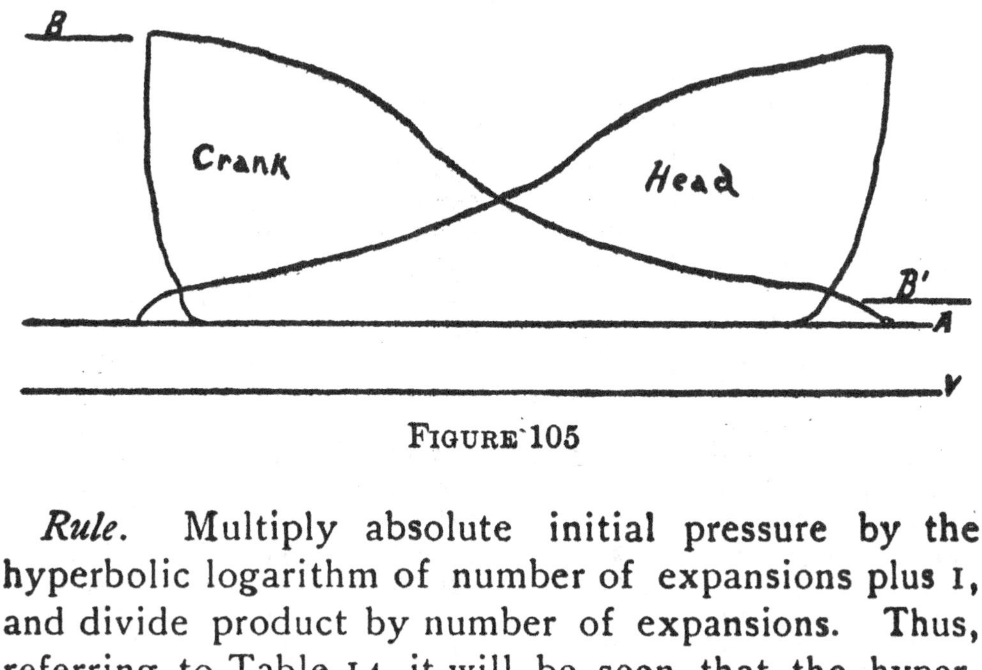

FIGURE 105

Rule. Multiply absolute initial pressure by the hyperbolic logarithm of number of expansions plus 1, and divide product by number of expansions. Thus, referring to Table 14, it will be seen that the hyperbolic logarithm of 3.65 is 1.2947, to which 1 must be added. Then $\frac{77 \times 2.2947}{3.65} = 48.4$ lbs., which is the absolute mean forward pressure. From this deduct the absolute back pressure, which is 16 lbs. or 1 lb. above atmosphere; thus, $48.4 - 16 = 32.4$ lbs. M. E. P.

Third. Find I. H. P.

Area of piston minus one-half area of rod ×

M. E. P. × piston speed in feet per minute, dividea by 33,000. Thus (the diameter of rod being 3 in.),

$$\frac{250.96 \times 32.4 \times 564}{33,000} = 138.9 \text{ I. H. P.}$$

The steam consumption per I. H. P. per hour may also be computed by means of Table 16, which was originally calculated by Mr. Thomson, and is based upon the following theory:

TABLE 16

T. P.	W.	T. P.	W.	T. P.	W.
3	117.30	13	466.57	23	798.10
3.5	135.75	13.5	483.43	23.5	814.39
4	153.88	14	500.22	24	830.64
4.5	171.94	14.5	517.07	24.5	846.96
5	186.75	15	533.85	25	863.25
5.5	207.60	15.5	550.64	25.5	879.49
6	225.24	16	567.36	26	895.70
6.5	242.97	16.5	584.10	26.5	911.86
7	260.54	17	600.78	27	927.99
7.5	278.06	17.5	617.40	27.5	944.07
8	295.44	18	633.96	28	960.12
8.5	312.80	18.5	650.46	28.5	976.27
9	330.03	19	666.90	29	992.38
9.5	347.27	19.5	683.38	29.5	1008.46
10	364.40	20	699.80	30	1024.50
10.5	381.57	20.5	716.27	30.5	1040.51
11	398.64	21	732.69	31	1056.48
11.5	415.73	21.5	749.06	31.5	1072.42
12	432.72	22	765.38	32	1088.32
12.5	449.69	22.5	781.76	32.5	1104.35

A horse-power = 33,000 ft. lbs. per minute, or 1,980,000 ft. lbs. per hour, or 1,980,000 × 12 = 23,760,-000 in. lbs. per hour, meaning that the same amount of energy required to lift 33,000 lbs. one foot high in one minute of time would lift 23,760,000 lbs. one inch high in one minute of time. Now, if an engine were driven by a fluid that weighed one pound per cubic inch, and the mean effective pressure of this fluid upon the piston was one pound per square inch, it would require 23,760,000 lbs. of the fluid per horse-power per hour. But, if in place of the heavier fluid we substitute pure distilled water, of which it requires 27.648 cu. in. to weigh one pound, the consumption per I. H. P. per hour will be considerably less; as, for instance, 23,760,000 ÷ 27.648 = 859,375 lbs., which would be the rate per hour of the water driven engine if the M. E. P. of the water was one pound per square inch and if the M. E. P. was increased to 20 lbs., then twenty times more power would be developed with the same volume of water, but the weight of water consumed per H. P. per hour would be proportionately less. Now, if the engine is driven by steam it will consume just as much less water in proportion as the water required to make the steam is less in volume than the steam used. Therefore if the above constant number, 859,375, be divided by the M. E. P. of any diagram and by the volume of the terminal pressure, the quotient will be the water (or steam) consumption per I. H. P. per hour.

Referring to Table 16, tne numbers in the W columns are the quotients obtained by dividing the constant, 859,375, by the volumes of the absolute pressures given in the columns under T. P. and which represent terminal pressures. The table is considerably abridged from the original, which was very full and complete, the pressures advancing by tenths of a pound from 3 lbs. to 60 lbs.; but it is seldom that in ordinary practice there is needed such accuracy. If at any time, however, a diagram should show a terminal pressure not given in the table, the correct factor for that pressure can be easily found by dividing the constant

859,375 by the relative volume of the pressure as found in Table 4 of the properties of saturated steam given in another chapter.

Referring again to Fig. 105, it is seen that the terminal pressure is 21 lbs. absolute, and by reference to Table 16 and glancing down column T. P. until 21 is reached, it will be seen that the number opposite in column W is 732.69. This number divided by the M. E. P. of the diagram Fig. 105, which is 32.4 lbs., gives 22.6 lbs. per I. H. P. per hour as the steam consumption. The rate thus found makes no allowance for clearance and compression, however, and these two very important items will be treated in a succeeding chapter, together with the method of correction for the above, viz., clearance and compression, as they enter largely into the steam economy of an engine.

Steam Consumption from Indicator Diagrams. In calculating the steam consumption of an engine, two very important factors must not be lost sight of, viz., clearance and compression. Especially is this the case in regard to clearance when there is little or no compression, for the reason that the steam required to fill the clearance space at each stroke of the engine is practically wasted, and all of it passes into the atmosphere or the condenser, as the case may be, without having done any useful work except to merely fill the space devoted to clearance. On the other hand, if the exhaust valve is closed before the piston completes the return stroke, the steam then remaining in the cylinder will be compressed into the clearance space and can be deducted from the total volume, which, without compression, would have been exhausted at the terminal pressure.

Figs. 106 and 107, which are reproductions of diagrams taken by the author while adjusting the valves on a 16 × 42 in. corliss engine, will serve to graphically illustrate this point. Fig. 106, which was the first one to be taken, shows no compression. The point of admission at A is plainly defined by the square corner at the extreme end of the stroke. The clearance of this engine is 4 per cent of the volume of the piston displacement. The engine being 16 in. bore by 42 in. stroke, the piston displacement is found by the following calculation: Area of piston, 201.06 sq. in. × stroke, 42 in. = 8444.52 cu. in. The volume of clearance space

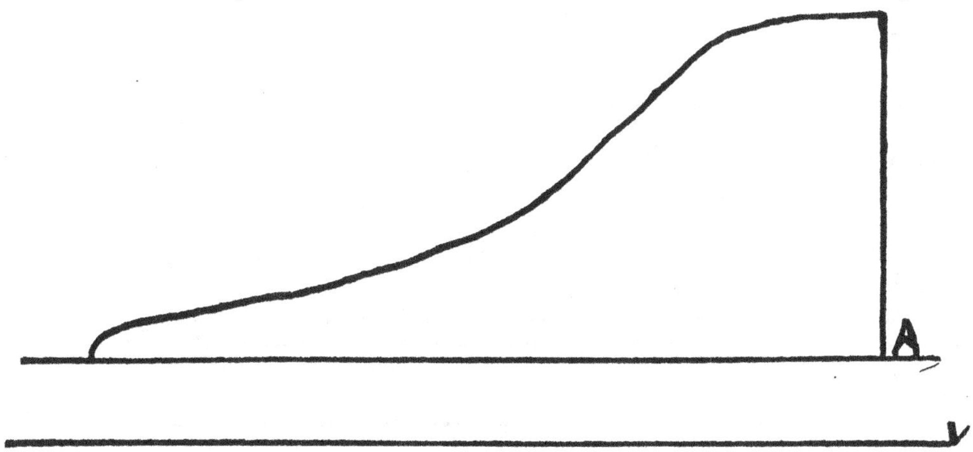

FIGURE 106

is equal to 8444.52 cu. in. × .04 = 337.78 cu. in., which divided by 1,728 = .195 cu. ft.

By reference to Fig. 107, taken after adjusting the valves for compression, it will be noticed that the steam is there compressed to 37 lbs., the compression curve beginning at C and ending at B. There is therefore compressed during each stroke a volume of steam equal to .195 cu. ft. at a pressure of 37 lbs. gauge, or 52 lbs. absolute.

One cubic foot of steam at 52 lbs. absolute pressure weighs .1243 lbs., and .195 cu. ft. will weigh .1243 × .195 = .0242 lbs.

The engine was running at 70 R. P. M., or 140 strokes per minute. Thus, according to Fig. 107, the total weight of steam compressed and doing useful work during one hour, and which without compression would have passed out through the exhaust pipe, is equal to .0242 × 140 × 60 = 203.28 lbs.

Now, in order to estimate the steam consumption of the above engine from diagram Fig. 106, it would be necessary to account for all the steam occupying not only the volume of the piston displacement at the end of the stroke, but the clearance as well, for the reason,

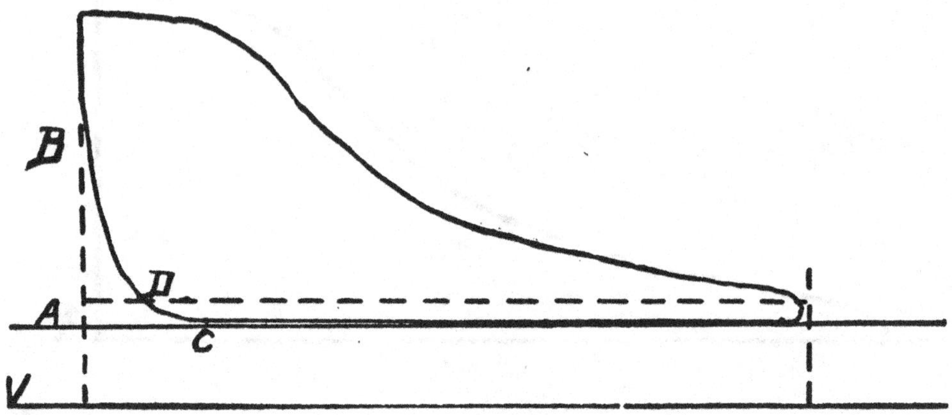

FIGURE 107

as before stated, that it would all be released before exhaust closure. This would equal 8444.52 cu. in. + 337.78 cu. in. = 8782.3 cu. in., which divided by 1,728 = 5.08 cu. ft. each stroke, or 10.16 cu. ft. each revolution.

The absolute terminal pressure of Fig. 106 is 20 lbs. One cubic foot of steam at this pressure weighs .0507 lbs., and the weight of steam consumed each revolution would therefore be 10.16 × .0507 = .515 lbs., which multiplied by 70 R. P. M. = 36.05 lbs. per minute, or 2,163 lbs. per hour. The horse-power developed by the engine at the time was 80. Therefore the steam consumption per I. H. P. per hour = 2,163 ÷ 80 = 27 lbs.

Referring again to Fig. 107, it will be remembered that the total weight of steam compressed during one hour was 203.28 lbs. The weight of steam consumed per hour, therefore, equals 2,163 – 203.28 = 1959.7 lbs.

Owing to compression, the work area of Fig. 107 is somewhat smaller than that of Fig. 106, amounting in fact to the area of the irregular figure enclosed between the points A, B and C. The work represented by this figure amounts to a very small proportion of the total work indicated by Fig. 106, still, in order to arrive at correct conclusions, it should be deducted therefrom.

Assuming the negative work to be equal to .55 horsepower, we have 80 – .55 = 79.45 I. H. P. as the work represented by Fig. 107. As the total weight of steam consumed in one hour was 1959.7 lbs., the steam consumption per I. H. P. per hour will be 1959.7 ÷ 79.45 = 24.67 lbs., a saving by compression of 2.33 lbs. per H. P. per hour, besides the great advantage of having a cushion of steam in contact with the piston at the termination of the stroke, thus bringing the moving parts of the engine to rest quietly without shock or jar.

The steam consumption may also be computed from the diagram, regardless of the dimensions of the cylinder or the horse power developed. The mean effective pressure and also the absolute terminal pressure must, however, be known. This method has been referred to, but in the computation therein made, no correction was made for clearance and compression.

Having reviewed these two factors at considerable length, it will now be in order to more fully explain the methods of treating diagrams when it is desired to make these corrections.

First, draw vertical lines C and D, Fig. 108, at each end of the diagram, and perpendicular to the atmospheric line. Draw line V, representing perfect vacuum, 14.7 lbs. below the atmospheric line, as indicated on the scale adapted to the diagram, which in this case is 50 lbs. to the inch. Continue the expansion from R, where release begins, until it intersects line D V, from which point the absolute terminal pressure can be measured.

Having ascertained the terminal pressure, which for Fig. 108 is 30 lbs., draw line D E, which may be called

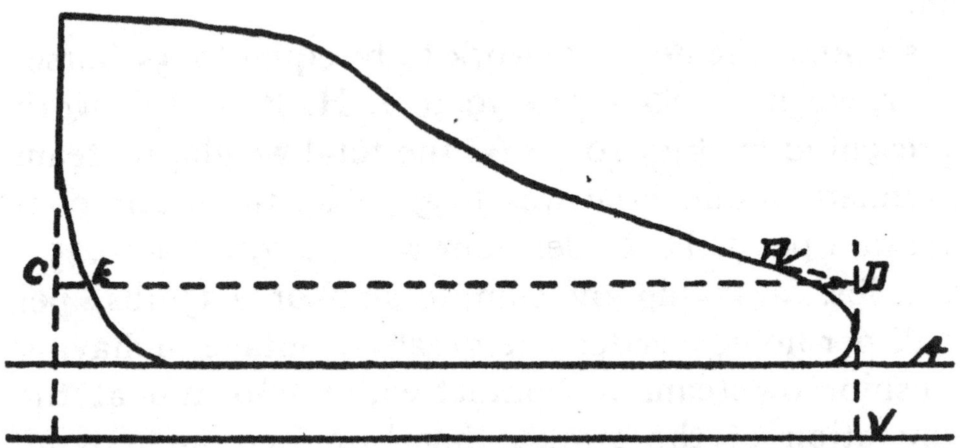

FIGURE 108

the consumption line for 30 lbs. The terminal being 30 lbs., refer to Table 16 and find in column W, opposite 30 in column T. P., the number 1,024.5. Divide this number by the M. E. P., which in Fig. 108 is 41 lbs., and the quotient, which is 24.99 lbs., is the uncorrected rate of steam consumption. This rate stands for the total consumption throughout the whole stroke represented on the diagram by the distance from D to C, which measures 3.25 in., but it is evident that there is a small portion of the return stroke, that indicated by the distance from E to C, during which the steam compressed in the clearance space should not be charged to the consumption rate, but should be deducted therefrom. In order to do this, multiply the uncorrected rate by the distance from D to E, which is 3⅛ in., or 3.125 in., and divide the product by the distance from D to C, 3¼ in., or 3.25 in. Thus, 24.99 × 3.125 ÷ 3.25 = 24.03 lbs., which is the corrected rate and represents a saving by compression of 24.99 − 24.03 = .96 lbs., or nearly 3.7 per cent.

In many cases the terminal pressure greatly exceeds the compression, an illustration of which is given in **Fig. 109.** It now becomes necessary to extend the

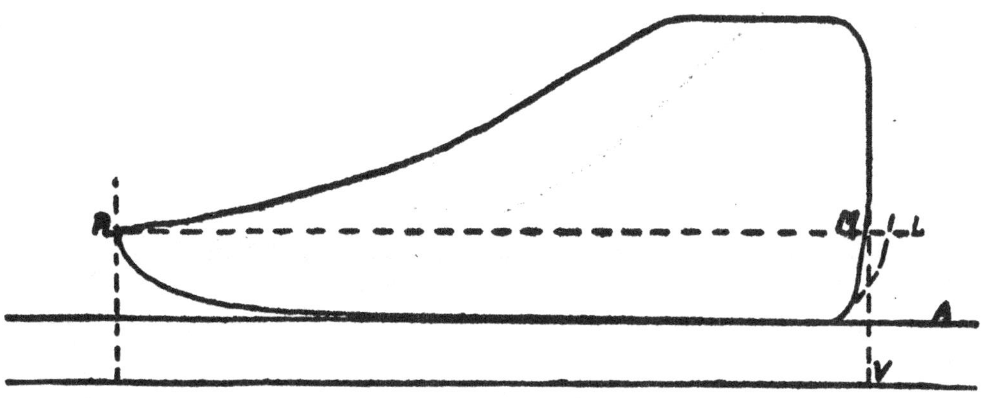

FIGURE 109

compression curve to L, a point equidistant from the vacuum line with the terminal at R. The consumption line R. L. now becomes longer than the stroke line R. M.; therefore the corrected rate will exceed the uncorrected rate by just so much; as, for instance, terminal pressure = 34 lbs. The factor, as per Table 16, = 1152.26, and the M. E. P. of the diagram is 47 lbs. Then, 1,152.26 ÷ 47 = 24.5 lbs., uncorrected rate; 24.5 × 3.125 in. (distance R. L.) ÷ 3 in. (distance R. M.) = 25.52 lbs., corrected rate, a loss of a little more than one pound, or about 4 per cent.

There is another class of diagrams very frequently encountered, in which the terminal pressure is considerably below the compression curve, and in order to compute the consumption rate by the above method it becomes necessary to continue the compression curve downwards until it meets the terminal, as illustrated at A, Fig. 110. R is the point of release, D A represents the consumption line, and D C the stroke. The terminal is 8.5 lbs., and the factor for that pressure, according to Table 16, is 312.8. Dividing this number

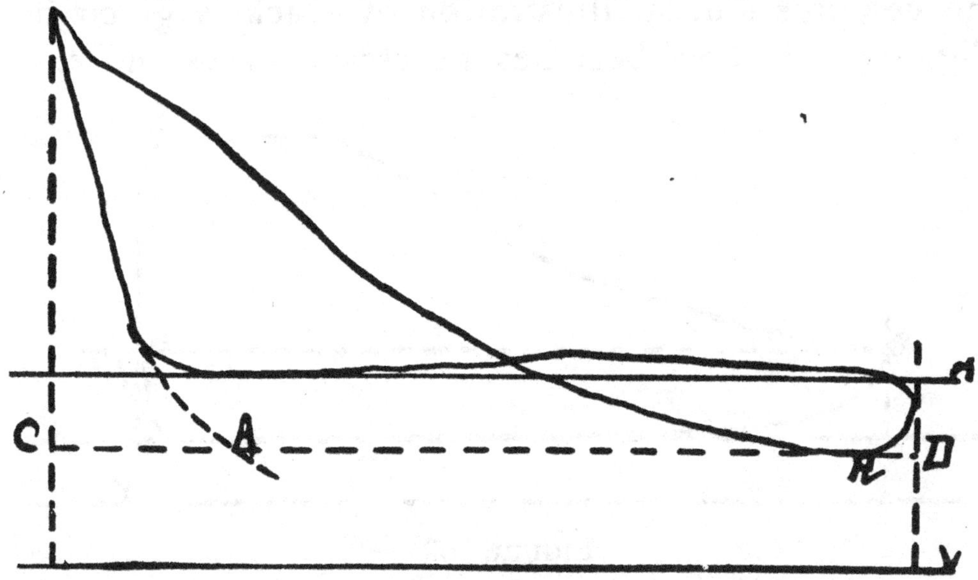

FIGURE 110

by the M. E. P., which was 7 lbs., gives 44.6 lbs. as the uncorrected rate. The distance D to A, where the compression curve intersects the consumption line, is 2.625 in., and the total length of the diagram C to D is 3.375 in. Then $44.6 \times 2.625 \div 3.375 = 35$ lbs. as the corrected rate.

Theoretical Clearance. The expansion and compression curves of a diagram are created by the expansion and compression of all the steam admitted during the stroke. This includes the steam in the clearance space as well as in the cylinder proper. It is evident, therefore, that the volume of the clearance is one of the factors controlling the form of these curves, and when the clearance is known a correct expansion or isothermal curve may be theoretically constructed, as will be explained later on. Also, if the actual curves, either expansion or compression, of a diagram assume an approximately correct form, the clearance, if not already known, may be determined theoretically from them; although too much confidence should not be put

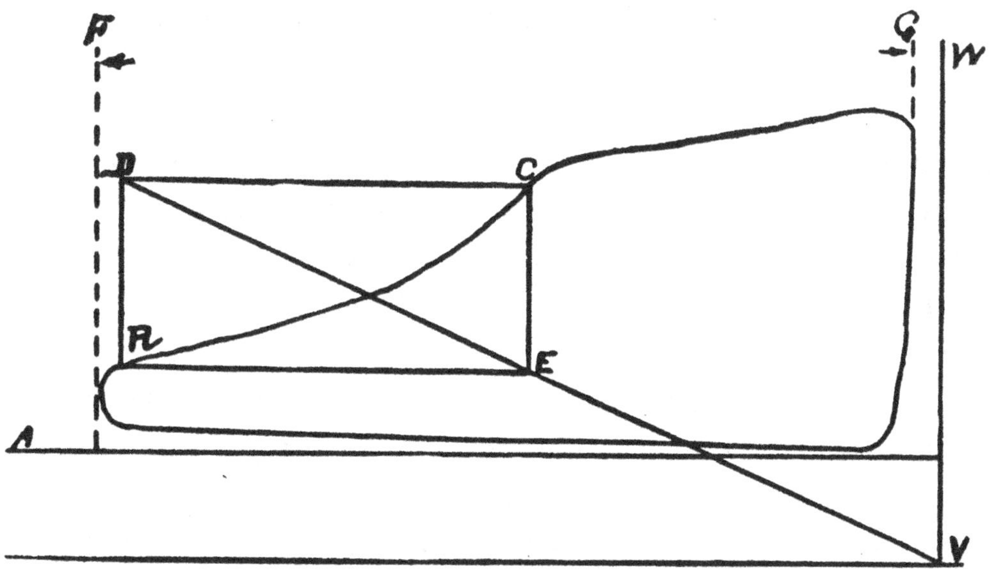

FIGURE 111

in the results, as they are liable to show either too little or too much clearance, generally the latter, especially if figured from the compression curve.

For the benefit of those who may desire to test this method of ascertaining the percentage of clearance of their engines, several illustrations will be given of its application to actual diagrams taken from engines in which the clearance was known.

Fig. 111 is from an engine in which the clearance was known to be 5 per cent. As compression cuts but a very small figure in this diagram, the expansion curve alone will be utilized for obtaining the theoretical clearance, and the process is as follows:

Select two points, C and R, in the curve as far apart as possible, but be sure that they are each within the limits of the true curve. Thus C is located just after cut-off takes place, and R is at a point just before release begins. From C draw line C D parallel with the atmospheric line. From D draw line D R, and from C draw line C E, both perpendicular to the atmospheric line. Then from R draw line R E, forming a rectangular parallelogram, C D R E, with two opposite corners, C and R, within the curve. Now through the other two corners, D and E, draw the diagonal D E, extending it downwards until it intersects the vacuum line V. From this point erect the vertical line V W, which is the theoretical clearance line.

To prove the result, proceed as follows: Measure the length of diagram from F to G, which in this case is 3.75 in., representing piston displacement. Next measure the distance from F to the clearance line V W, which is 3.91 in., representing piston displacement with volume of clearance added. Then $3.91 - 3.75 = .16$, which represents volume of clearance; and $.16 \times 100 \div 3.75 = 4.3$ per cent, which is approximately near the actual clearance, which, as before stated, was 5 per cent.

The Theoretical Expansion Curve. According to Boyle's law the volume of all elastic gases is inversely as their pressures, and steam, being a gas, conforms substantially to this law; although the expansion curves of indicator diagrams are affected more or less by the loss of heat transmitted through the cylinder walls, and by the change in the temperature of the steam produced by the changes in pressure during the progress of the stroke. The pressure generally falls more rapidly during the first part of the stroke, and less rapidly during the last portion than it should in order to conform strictly to the above law, and the terminal pressure usually is greater than it should be to agree with the ratio of expansion. But this fullness of the expansion curve of the diagram near the end compensates in a measure for the too rapid fall near

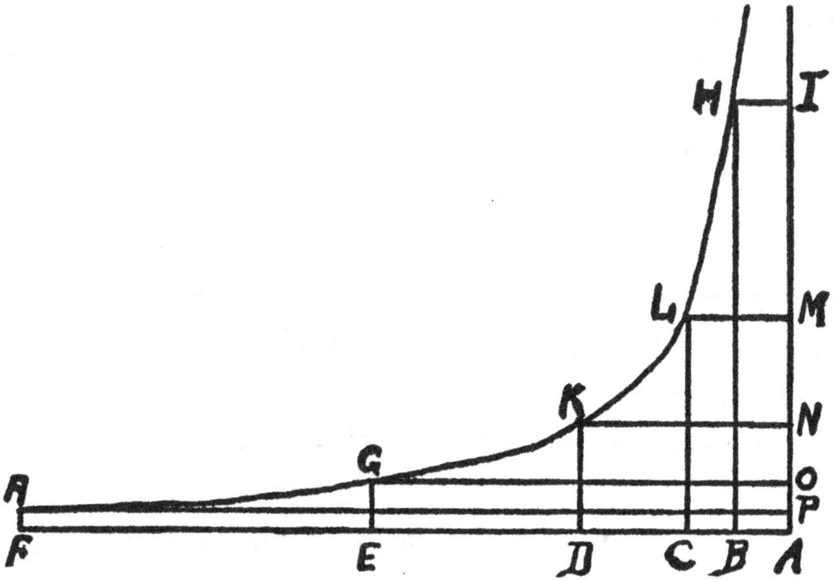

FIGURE 112

the beginning of the stroke. Therefore, if the engine is in fairly good condition, with the valves properly adjusted and not leaking, and the piston rings are steam-tight, it may be assumed that the expansion of the steam in the cylinder takes place according to Boyle's law, and it is found that the expansion curve drawn by the indicator practically coincides with a hyperbolic curve constructed according to that law.

Fig. 112 graphically illustrates the application of the hyperbolic law to the expansion of gases. The horizontal lines represent volumes and the vertical lines represent pressures. The base line, A F, represents the full stroke of a piston in the cylinder of an engine, and the vertical line A I represents the pressure of the steam at the commencement of the stroke.

Suppose there is no clearance and that the steam has been admitted up to point H when it is cut off. The rectangle A B H I is the product of the pressure multiplied by the volume of the steam thus admitted. When the piston has traveled from A to C the volume of the steam has been doubled and the pressure C L has been reduced to just one-half what it was at A I, but the area of the rectangle A C L M is equal to the area of the initial rectangle, and, as before, is the product of the pressure C L multiplied by the volume A C. As the piston travels still farther, as from A to D, the steam is expanded to four volumes, while the pressure at D K will only be one-fourth that of the initial pressure; but the new rectangle A D K N is still equal in area to either of the others, A B H I or A C L M.

The same law applies to each of the remaining rectangles; A E G O representing five volumes and one-fifth of the initial pressure, and A F R P representing six times the initial volume and one-sixth of the initial pressure, but each having the same area as the initial rectangle A B H I. Now, the area of the rectangle A B H I represents the work done by the steam up to the point of cut-off, and the area of the hyperbolic figure enclosed by the lines B H R F represents the work done by the expansion of the steam after cut-off occurs. This area and the amount of work it represents may be computed by means of the known relations of hyperbolic surfaces with their base lines; as, for instance, if the base lines A B, A C, A D, etc., extend in geometrical ratio, as 1, 2, 4, 8, 16, etc., the successive areas, B H L C, B H K D, B H G E, etc., increase in arithmetical ratio, as 1, 2, 3, 4, etc.

On the principles of common logarithms, which represent in arithmetical ratio natural numbers in geometrical ratio, tables of hyperbolic logarithms have been computed for the purpose of facilitating the calculation of areas of work due to different degrees of expansion. Such a table is given elsewhere in this book, and the method of calculating the M. E. P. by this means is described

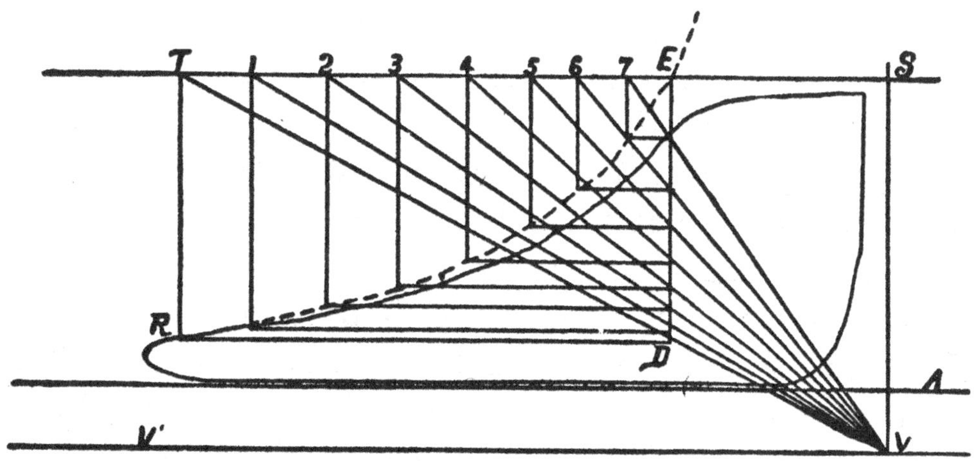

FIGURE 113

A theoretical curve may be constructed conjointly with the actual expansion curve of a diagram by first locating the clearance and vacuum lines and then pursuing the method illustrated by Fig. 113. A curve so produced is called an isothermal curve, meaning a curve of the same temperature.

Referring to Fig. 113, suppose, first, that it is desired to ascertain how near the expansion curve of the diagram coincides with the isothermal curve, at or near the point of cut-off. Select point R near where release begins, but still well within the expansion curve. From this point draw the vertical line, R T, parallel with the clearance line, V S. Then draw the horizontal line, S T, parallel with the atmospheric line, and at such a height above it as will equal the boiler pressure as measured by the scale adapted to the diagram; such measurement to be made from the atmospheric line to correspond with the gauge pressure. From T draw the diagonal T V, and from R draw the horizontal line R D parallel with the atmospheric line. From D, where this line intersects T V, erect the perpendicular D E, thus forming the parallelogram R D E T, and as line T V passes through two of its opposite angles and meets the junction of the clearance and vacuum lines, the other two angles, R and E, will be in the theoretical curve, and R being the starting point, it is obvious that this curve must pass through E, which would be the theoretical point of cut-off on the steam line S T.

Two important points in the theoretical curve have now been located, viz., E as the cut-off, and R as the point of release. In order to obtain intermediate points, draw any desired number of lines downward from points in S T, as 1, 2, 3, 4, 5, etc., and continue them downwards far enough to be sure that they will meet the intended curve, and from the same points in S T draw diagonals 1 V, 2 V, 3 V, 4 V, 5 V, etc., all to converge accurately at V. From the intersection of these diagonals with D E draw horizontal lines parallel with V V′, and the points of junction of these lines with the vertical lines will be points in the theoretical curve. It will now be an easy matter to trace the curve through these points. If, on the other hand, it be desired to compare the curves toward the exhaust end of the diagram, draw lines E D and E T, Fig. 114, also T R, locating R near where release commences, after which draw line R D, completing the parallelogram E T R D, fixing R as a point in the theoretical curve started at E. After drawing the diagonal T V, proceed in the same manner as before to locate the intermediate points.

It will be observed that in order to ascertain the performance of the steam near the beginning of the stroke, the starting point of the isothermal curve must be near the point of release, and conversely, if the starting point of the curve is located near the point of

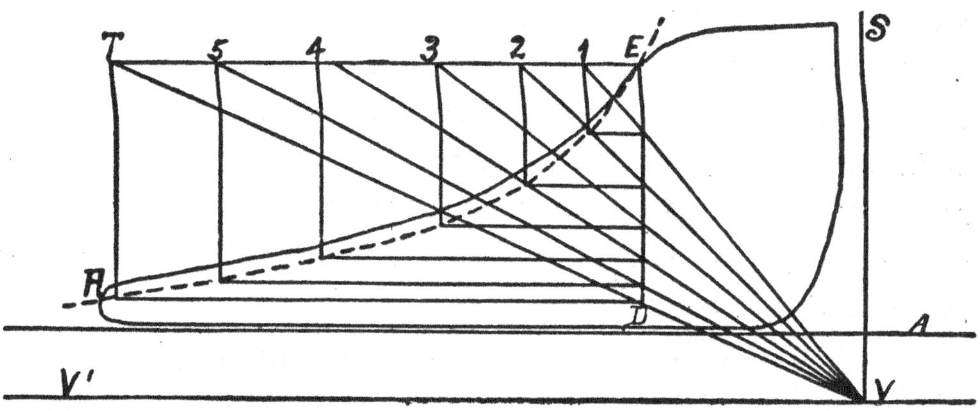

FIGURE 114

cut-off and coincident with the actual curve, the test will apply towards the end of the stroke. It is not to be expected that the expansion curve of any diagram taken in practice will conform strictly to the lines of the isothermal curve, especially towards the latter end of the stroke, owing to the reëvaporation of water resulting from the condensation of steam which was retained in the cylinder by the closing of the exhaust valve. This reëvaporation commences just as soon as the temperature of the steam owing to reduction of pressure due to expansion, falls below the temperature of the cylinder walls, and it continues at an increasing rate until release occurs. The tendency of this reëvaporation or generation of steam within the cylinder during the latter portion of the stroke is to raise the terminal pressure considerably above what it would be if true isothermal expansion took place. The terminal pressure may also be augmented by a leaky steam valve, while, on the other hand, a leaky piston would cause a lowering of the terminal and an increase in the back pressure.

The Adiabatic Curve. If it were possible to so protect or insulate the cylinder of a steam engine that there would be absolutely no transmission of heat either to or from the steam during expansion, a true adiabatic curve or "curve of no transmission" might be obtained. The closer the actual expansion curve of a diagram conforms to such a curve, the higher will be the efficiency of the engine as a machine for converting heat into work.

Fig. 115 illustrates a method of figuring a curve which, while not strictly adiabatic, will be near enough for all practical purposes, while at the same time it will give the student an opportunity to study the laws governing the expansion of saturated steam.

To draw the curve, first locate the clearance and vacuum lines V S and V V′. Next locate point R in the expansion curve near where release begins, making this the starting point, and also the point of coincidence of the expansion curve with the adiabatic curve.

The other points in the curve are located from the volumes of steam at different pressures during expansion; the pressures being measured from the line of perfect vacuum, and the volumes from the clearance line.

The absolute pressure at R, Fig. 115, is 26 lbs.

From point R erect the perpendicular R T. Also draw horizontal line R 26 parallel with the vacuum line and at a height equal to 26 lbs. above vacuum line V V′, as shown by the scale, which in this case was 40. The length of line R 26, measured from R to the clearance line, is $3\frac{1}{16}$ in., or 3.0625 in. By reference to Table 4 it will be seen that the volume of steam at 26 lbs. absolute, as compared with water at 39°, is 962. Now, if the length of line R 26 be divided by this volume, and the quotient multiplied by each of the volumes of the other pressures represented at points 30, 35, 40,

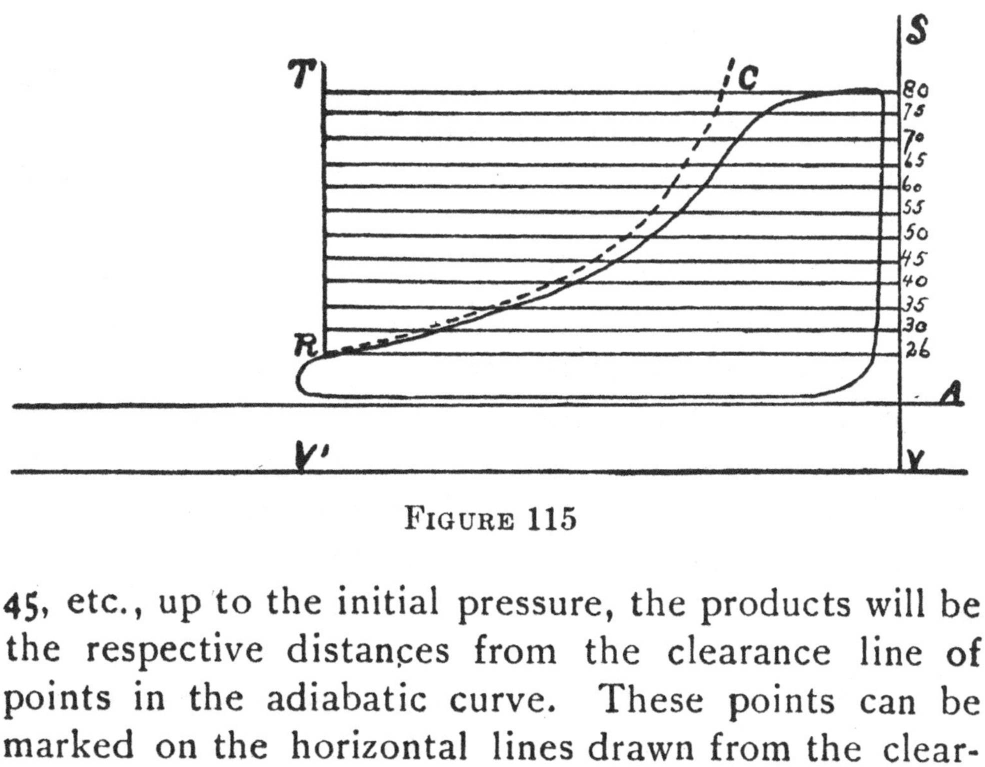

FIGURE 115

45, etc., up to the initial pressure, the products will be the respective distances from the clearance line of points in the adiabatic curve. These points can be marked on the horizontal lines drawn from the clearance line to line R T.

Starting with line R 26, it has been noted that its length is 3.0625 in., and that the volume was 962. 3.0625 ÷ 962 = .003. Then the volume of steam at 30 lbs. is 841, which being multiplied by .003 = 2.5 in., the length of line 30. Next, the volume at 35 lbs. = 728.

Multiplying this volume by .003 = 2.1 in., length of line 35, and so in like manner for each of the other points.

The process involves considerable figuring and careful and accurate measurements, which should be made with a steel rule with decimal graduations. It is not expected that the cut Fig. 115 will be found accurate enough in its measurements to serve as a standard; it being intended only to serve as an illustration of the process. The diagram from which the illustration was drawn was taken from a 600 H. P. engine situated some 200 ft. from the boilers, and there was a considerable cooling of the steam by the time it reached the engine, the effect of which is apparent. The curve produced by the measurements is shown by the broken line. The process can be applied to any diagram.

Power Calculations. The area of the piston (minus one-half the area of rod) multiplied by the M. E. P., as shown by the diagram, and this product multiplied by the number of feet traveled by the piston per minute (piston speed), will give the number of foot-pounds of work done by the engine each minute, and if this product be divided by 33,000, the quotient will be the indicated horse-power (I. H. P.) developed by the engine.

Therefore one of the first requisites in power calculations is to ascertain the M. E. P. Beginning with the most simple, though only approximately correct, method of obtaining the average pressure, as illustrated by Fig. 116, draw line A B touching at A and cutting the diagram in such manner that the space D above it will equal in area spaces C and E taken together, as nearly as can be estimated by the eye.

Then with the scale measure the pressure along the line F G at the middle of the diagram, which will be the M. E. P

The process is based upon the theory that the average width of any tapering figure is its width at the middle of its length. This method should not be relied upon as accurate, but is convenient at times when it is desired to make a rough estimate of the horse-power of an engine.

Figuring the M. E. P. by Ordinates. This is a very common method and one which can be relied upon to

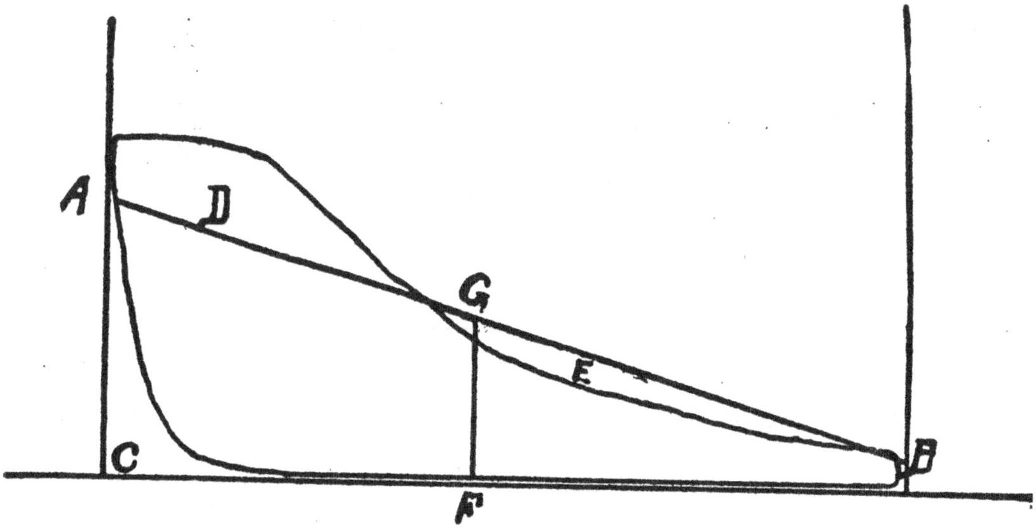

FIGURE 116

give accurate results, provided care is exercised in its use.

The process consists in drawing any convenient number of vertical lines perpendicular to the atmospheric line across the face of the diagram, spacing them equally, with the exception of the two end spaces, which should be one-half the width of the others, for the reason that the ordinates stand for the centers of equal spaces, as, for instance, line 1, Fig. 117, stands for that portion of the diagram from the end to the middle of the space between it and line 2. Again, line 2 stands for the remaining half of the second space and the first half of the third, and so on. This is an important matter, and should be thoroughly understood, because if the spaces are all made of equal width, and measurements are taken on the ordinates, the results will be incorrect, especially in the case of high initial pressure and early cut-off, following which the steam undergoes great changes.

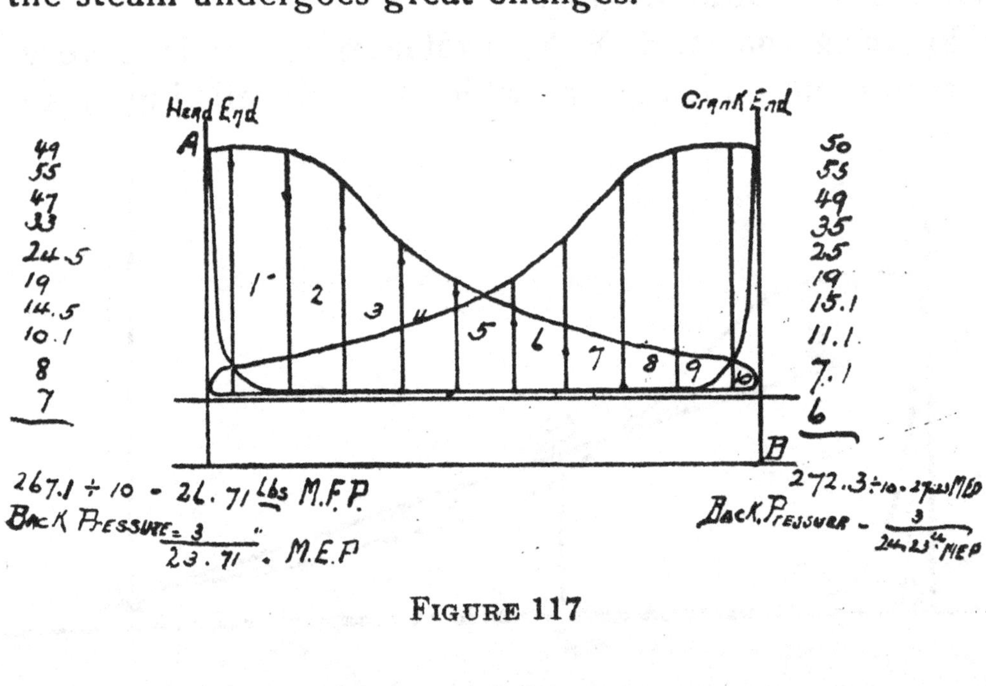

FIGURE 117

If the spaces are all made equal, the measurements will require to be taken in the middle of them, and errors are liable to occur, whereas, if spaced as before described, the measurements can be made on the ordinates, which is much more convenient and will insure correct results. Any number of ordinates can be drawn, but ten is the most convenient and is amply sufficient, except in case the diagram is excessively long. For spacing the ordinates, dividers may be used, or a parallel ruler may be procured from the makers of the indicator; but one of the most con venient and easily procurable instruments for this purpose is a common two-foot rule, and the method of using it is illustrated in Fig. 117.

First draw vertical lines at each end of the diagram, perpendicular to the atmospheric line and extending downwards to the vacuum line, or below it, if necessary, in order to have a point on which to lay the rule. In Fig. 117, points A and B are found to be the most convenient. Now lay the rule diagonally across the diagram, touching at A and B, and the distance will be found to be $3\frac{3}{4}$ in., or 60 sixteenths.

Suppose it be desired to draw 10 ordinates. Divide 60 by 10, which will give 6 sixteenths, or $\frac{3}{8}$ in., as the width of the spaces, but as the two end spaces are to be one-half the width of the others, there will be 11 spaces altogether, the two outer ones having a width equal to one-half of $\frac{3}{8}$, or $\frac{3}{16}$. Now apply the rule again in the same manner, touching at points A and B, and with a sharp pointed pencil begin at A and mark the location of the first ordinate according to the rule, at a distance of $\frac{3}{16}$ from the end. Then $\frac{3}{8}$ from this mark make another one, which will locate the second ordinate, and proceed in like manner to locate the others. The last two or three marks generally come below the diagram, and if the diagram be taken from a condensing engine it may be necessary to tack it on to a larger sheet of paper in order to get these points. Having correctly located the ordinates, they may now be drawn perpendicular to the atmospheric line or vacuum line, either of which will answer.

It should be noted that, owing to the diagonal position of the rule with relation to the atmospheric line, the spaces are not of the actual width as described by the rule, but this is unimportant, so long as they are of a uniform width. This method can be applied to any diagram, no matter what its length may be, and point B may be located at any distance below the atmospheric or vacuum lines, wherever it is the most convenient for the subdivisions on the rule, sixteenths, eighths, etc., so long as it is in line with the end of the diagram. Having thus drawn the ordinates, the M. E. P. may be found by measuring the pressure expressed by each one, using for this purpose the scale adapted to the spring used, adding all together and dividing by the number of ordinates which will give the average pressure.

Referring to Fig. 117, begin with ordinate No. 1 on the diagram, from the head end of the cylinder. In this case a 40 spring was used. Lay the scale on the ordinate with the zero mark where it intersects the compression curve. The pressure is seen to be 49 lbs. Set this down at that end of the card and measure the pressure along ordinate No. 2, which is 55 lbs. Proceed in this manner to measure all the ordinates, placing the resulting figures in a column, after which add them together and divide by 10. The result is 26 71 lbs., which is the mean forward pressure (M. F. P.). To obtain the mean effective pressure, deduct the back pressure, which is represented by the distance of the exhaust line of the diagram above the atmospheric line in a non-condensing engine, and in a condensing engine the back pressure is measured from the line of perfect vacuum, 14.7 lbs., according to the scale below the atmospheric line.

In Fig. 117 the back pressure is found to be 3 lbs. Therefore the M. E. P. of the head end will be 26.71 − 3 = 23.71 lbs. On the crank end the M. F. P. is 27.23 lbs., and 27.23 − 3 = 24.23 lbs. = M. E. P. The average effective pressure on the piston, therefore, will be $23.71 + 24.23 \div 2 = 23.97$ lbs.

Unless great care is exercised in the measurements, errors are liable to occur in applying this method, especially with scales representing high pressures, as 60, 80, etc. The most convenient and reliable method is to take a narrow strip of paper of sufficient length, and starting at one end, apply its edge to each ordinate in succession and mark their lengths on it consecutively with the point of a knife blade or a sharp pencil. Having thus marked on the paper the total length of all the ordinates, ascertain the number of inches and fractions of an inch thereon, the fractions to be expressed decimally, and divide by the number of ordinates. The quotient will be the average height of the diagram, and as the scale expresses the number of pounds pressure for each inch or fraction of an inch in height, if the average height of the diagram be multiplied by the number of the scale, the product will be the M. F. P.

Referring again to Fig. 117, if the lengths of the ordinates drawn on the head end diagram be measured, their sum will be found to be $6\frac{8}{12}$ or 6.666 in. Dividing this by 10 gives .666 in. as the average height. The mean forward pressure will then be as follows: $.666 \times 40 = 26.64$ lbs., or practically the same as found by the other method.

Fig. 118 illustrates a type of diagram frequently met with, and one which requires somewhat different treatment in estimating the power developed. It will be noticed that, owing to light load and early cut-off, the expansion curve drops considerably below the atmospheric line, notwithstanding that the engine from which this diagram was taken is a non-condensing engine. When release occurs at R, and the exhaust side of the piston is exposed to the atmosphere, the pressure immediately rises to a point equal to, or slightly above, that of the atmosphere.

Fig. 118 was taken during a series of experiments made by the author for the purpose of ascertaining the friction of shafting and machinery, and the engine it was obtained from is a Buckeye 24 x 48 in. The boiler pressure at the time was only 40 lbs., and a No. 20 spring was used. The ordinates are drawn accord-

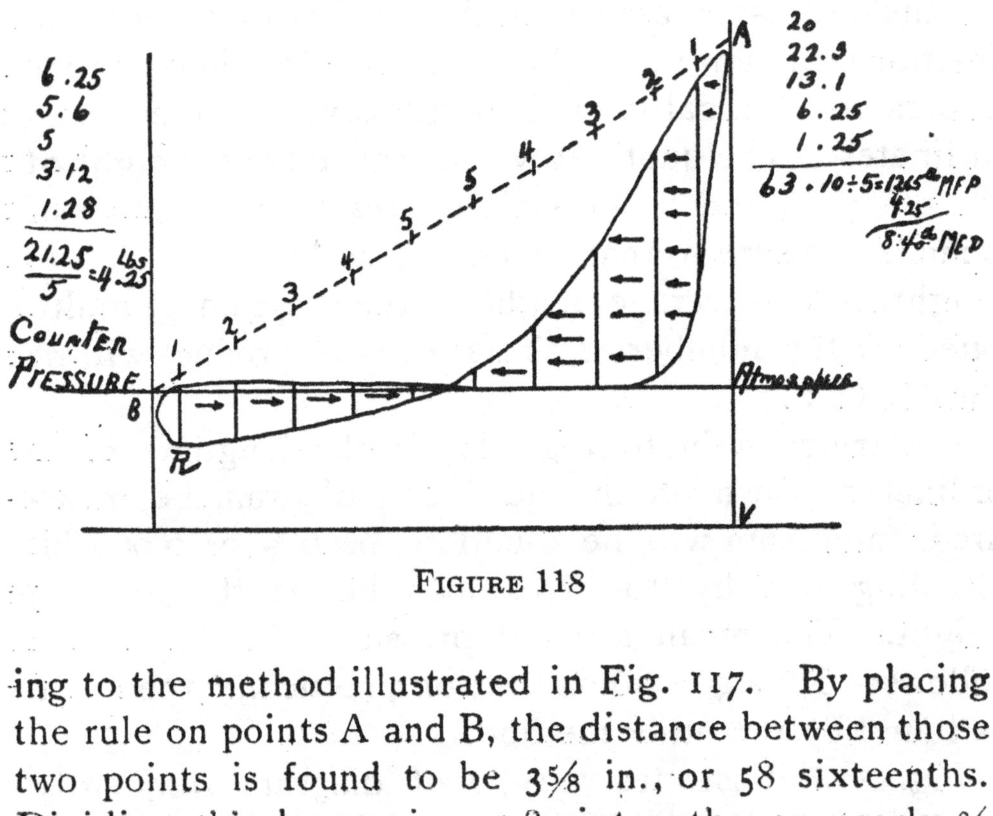

FIGURE 118

ing to the method illustrated in Fig. 117. By placing the rule on points A and B, the distance between those two points is found to be 3⅝ in., or 58 sixteenths. Dividing this by 10 gives 5.8 sixteenths, or nearly ⅜ in., as the width of the spaces; the two end spaces being one-half of this, or $\frac{3}{16}$ in. wide. The first five ordinates, counting from A, express forward pressure, represented by the arrows. The remaining five ordinates, counting from B, express counter or back pressure, represented by the arrows pointing in the opposite direction. Measuring the pressures along the first five ordinates, and adding them together, gives 63.1 lbs., which divided by 5 gives 12.65 lbs. as the mean forward pressure (M. F. P.).

Then figuring up the counter pressure in the same manner on the other five ordinates, beginning at B, the result is 4.25 lbs. The M. E. P., therefore, will be $12.65 - 4.25 = 8.4$ lbs.

Obtaining the M. E. P. with the Planimeter. The area of the diagram represents the actual work done by the steam acting upon the piston. In a non-condensing engine the lower or exhaust line of the diagram must be either coincident with or slightly above the atmospheric line in order to express positive work. Any deviation of this line, either above or below the atmospheric line, represents counter pressure, the amount of which may be ascertained by measurements with the scale, and should be deducted from the mean forward pressure.

On the other hand, the exhaust line of a diagram from a condensing engine falls more or less below the atmospheric line, according to the degree of vacuum maintained, and the nearer this line approaches the line of perfect vacuum, as drawn by the scale, 14.7 lbs. below the atmospheric line, the less will be the counter pressure, which in this case is expressed by the distance the exhaust line is above that of perfect vacuum.

The prime requisite, therefore, in making power calculations from indicator diagrams is to obtain the average height or width of the diagram, supposing it were reduced to a plain parallelogram instead of the irregular figure which it is.

The planimeter, Fig. 119, is an instrument which will accurately measure the area of any plane surface, no matter how irregular the outline or boundary line is, and it is particularly adapted for measuring the areas of indicator diagrams, and in cases where there are many diagrams to work up, it is a very convenient instrument and saves much time and mental effort. In fact, the planimeter has of late years become an almost indispensable adjunct of the indicator. It shows at once the area of the diagram in square inches and decimal fractions of a square inch, and when the area is thus known it is an easy matter to obtain the average height by simply dividing the area in inches by the length of the diagram in inches. Having ascertained the average height of the diagram in inches or fractions of an inch, the mean or average pressure is found by multiplying the height by the scale. Or the process may be made still more simple by first multiplying the area, as shown by the planimeter in square inches and decimals of an inch, by the scale, and dividing the product by the length of the diagram in inches. The result will be the same as before, and troublesome fractions will be avoided.

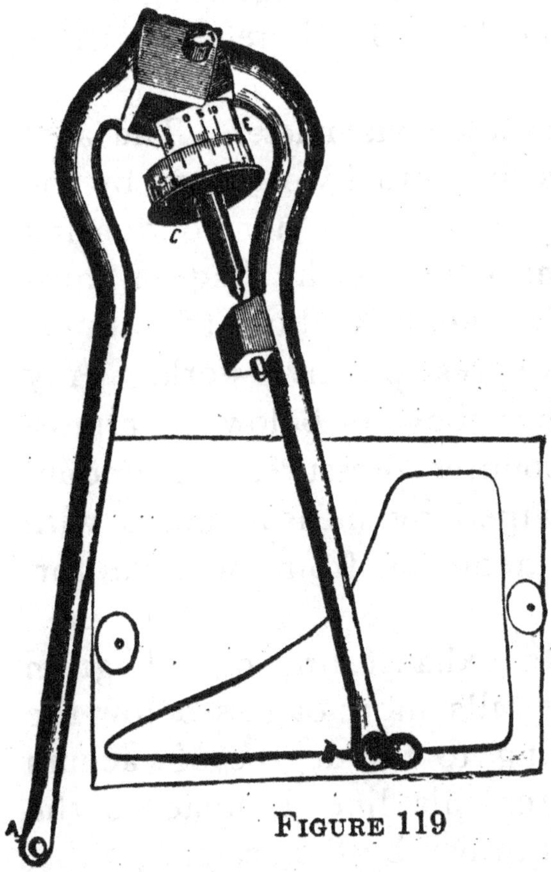

FIGURE 119

THE INDICATOR

Questions

317. Who invented the indicator, and for what purpose did he apply it to his engine?

318. What are the principles governing the action of the indicator?

319. What will a truthful diagram from a steam engine cylinder show?

320. Describe in general terms the construction of an indicator.

321. Does the steam act upon both sides of the indicator piston?

322. What does the atmospheric line show?

323. Is this line important in the study of indicator diagrams?

324. Where should the line of back pressure appear in a diagram from a non-condensing engine?

325. What controls the length of stroke of the indicator piston?

326. What does the number on the spring mean?

327. Should the pencil fall below the atmospheric line in a diagram from a locomotive?

328. What is the most practical device for reducing the motion of the cross head to correspond with the motion of the indicator drum?

329. What are the main requirements in indicator connections?

330. What should be done with these pipes before attaching the indicator?

331. What regulates the height of the diagram?

332. What is a convenient rule to be observed in the selection of the spring?

333. What governs the length of the diagram?

334. Describe the best method of tracing the atmospheric line.

335. What data should be noted on the diagram as soon as taken?

336. By what means may the taking of indicator diagrams from locomotives be greatly facilitated?

337. What is absolute pressure?

338. What is gauge pressure?

339. What is initial pressure?

340. What is terminal pressure and how may it be ascertained theoretically?

341. What is back pressure?

342. What is absolute back pressure?

343. What is meant by ratio of expansion?

344. What does the term wire-drawing mean when applied to an indicator diagram?

345. What is condenser pressure?

346. What does the term vacuum imply?

347. What is absolute zero?

348. What is meant by the term piston displacement?

349. What is piston clearance?

350. What is steam clearance?

351. What is a horse-power?

352. What is meant by piston speed?

353. Define Boyle's law of expanding gases.

354. What is an adiabatic curve?

355. What is an isothermal curve?

356. What is the first law of thermodynamics?

357. What is the unit of work?

358. Define the first law of motion?

359. What is momentum?

360. What is the maximum theoretical duty of steam?

361. What is meant by the term steam efficiency?

362. How may the term engine efficiency be defined?

363. What are common logarithms?

364. What are hyperbolic logarithms, and how are they found?

365. What are ordinates as applied to an indicator diagram?

366. What is an eccentric?

367. What is meant by the throw of an eccentric?

368. What is meant by position of the eccentric?

369. What is angular advance?

370. What is meant by the expression, steam consumption of an engine?

371. What effect has back pressure upon the work of an engine?

372. What relation should the steam line of a diagram bear to the atmospheric line?

373. In calculations for steam consumption what two important factors must be considered?

374. How is the piston displacement of an engine ascertained?

375. What do the expansion and compression curves of a diagram show?

376. Is steam a gas?

377. What effect does reëvaporation have upon the expansion curve?

378. How is the horse-power of an engine calculated?

379. What is meant by the expression M. E. P.?

380. What is a planimeter?

CHAPTER VIII

COMPOUND LOCOMOTIVES

The principal object in compounding locomotives is to effect economy in fuel, and this economy is due to the fact that with the compound engine the steam may be expanded to a much lower pressure than is possible with the simple engine, before it is allowed to exhaust into the atmosphere. Another source of economy in compounding the cylinders of any steam engine, stationary or locomotive, is the prevention of that excessive condensation which is sure to result when steam at a high pressure is admitted to a cylinder, the walls of which are at a comparatively low temperature at the moment of admission, and this takes place at each stroke of the simple engine; as, for instance, assume the initial pressure to be 195 lbs., and the pressure at release to be 8 lbs. The temperature of steam at 195 lbs. is 385°, and at 8 lbs. pressure the temperature is 235°. This drop of 150° in the temperature during each stroke of the engine, tends to cool the walls of the cylinder, which will be warmed again by the next admission of steam. A large amount of heat is thus being continually absorbed by the cylinder walls, and there is also a constant loss caused by condensation.

In the compound locomotive the expansion of the steam is divided between two cylinders, proportioned in such a way that the amount of work done in each will be the same.

Various types of compound locomotives have been designed and built by eminent engineers in this country and in Europe, and while, as before stated, the main object in compounding is to utilize as much of the tremendous energy stored in the coal as it is possible to utilize, still there are other important problems to be solved in the design and operation of compound locomotives, not the least of which is to so proportion the cylinders, especially of a cross compound, that there will be an equal distribution of power on each side of the engine, or, in other words, that the engine will be balanced. Another problem that has been constantly before the purchaser and the builder of compound locomotives, is that of keeping the number of parts down to as low a figure as possible, and thus to produce a machine that will use steam on the compound principle, and yet at the same time eliminate as far as possible the liability of additional expense for repairs that has always been connected with the compound as compared with the simple engine.

The progress along these lines has been slow, but there has been a marked development in the right direction, and there is no doubt that the compound locomotive has come to stay, and that eventually it will become the standard type. It therefore behooves engine men (engineers and firemen) to study them, and endeavor to familiarize themselves with their construction and operation.

There are in use at the present time, in this country, four separate and distinct types of compound locomotives, each having its peculiar features. First, there is the Vauclain compound. This is a four cylinder engine, having two cylinders on each side of the engine. One of these cylinders is a high-pressure and the other a low-pressure cylinder, one being located directly above the other.

Second, the balanced compound, a four cylinder engine, the two high-pressure cylinders being located under the center of the smoke arch, between the frames, and the two low-pressure cylinders on the outside.

Third, the tandem compound, a four cylinder engine, having one high and one low-pressure cylinder on each side, these cylinders being in line with each other, and served by one piston rod, thus bringing all the strains in direct line also.

Fourth, the cross compound, a two cylinder engine, having the high-pressure cylinder on one side, and the low-pressure cylinder on the opposite side, the diameter or bore of the cylinders being proportioned in such manner that an equal amount of power will be developed on both sides. This ratio is generally one to three, that is, the area of the low-pressure piston is about three times that of the high, for the reason that the initial pressure of the steam admitted to the low-pressure cylinder is greatly reduced below the point at which it entered the high-pressure cylinder, and requires a larger area of piston to act upon in order to produce the same amount of power that it did in the high-pressure cylinder. These four forms of compound locomotives will be taken up, and each discussed in its regular order. The same valve gear is used upon compound locomotives as upon simple engines or those in which there is but single expansion.

The Vauclain compound locomotive is the invention of Mr. Samuel M. Vauclain of the Baldwin Locomotive Works, and the following description of this system of compounding has been mainly furnished by the Baldwin Locomotive Works of Philadelphia, Pa.

In designing the Vauclain system of compound locomotives, the aim has been:

1. To produce a compound locomotive of the greatest efficiency, with the utmost simplicity of parts and the least possible deviation from existing practice. To realize the maximum economy of fuel and water.

2. To develop the same amount of power on each side of the locomotive, and avoid the racking of machinery resulting from unequal distribution of power.

3. To insure at least as great efficiency in every respect as in a single-expansion locomotive of similar weight and type.

4. To insure the least possible difference in cost of repairs.

5. To insure the least possible departure from the method of handling single-expansion locomotives; to apply equally to passenger or freight locomotives for all gauges of track, and to withstand the rough usage incidental to ordinary railroad service.

The principal features of construction are as follows:

Cylinders. The cylinders consist of one high-pressure and one low-pressure for each side, the ratio of the volumes being as nearly three to one as the employment of convenient measurements will allow. They are cast in one piece with the valve-chamber and saddle, the cylinders being in the same vertical plane, and as close together as they can be with adequate walls between them.

Where the front rails of the frames are single bars, the high-pressure cylinder is usually put on top, as shown in Fig 120, but when the front rails of frames are double, the low-pressure cylinder is usually on top, as shown in Fig. 121.

The former (Fig. 120) is used in "eight-wheel" or

American type passenger locomotives, and in "ten-wheeled" locomotives, while the latter (Fig. 121) is used in Mogul, Consolidation and Decapod locomotives; for the various other classes of locomotives the most suitable arrangement is determined by the style of frames.

Fig. 122 shows the arrangement of the cylinders in relation to the valve.

The valve employed to distribute the steam to the

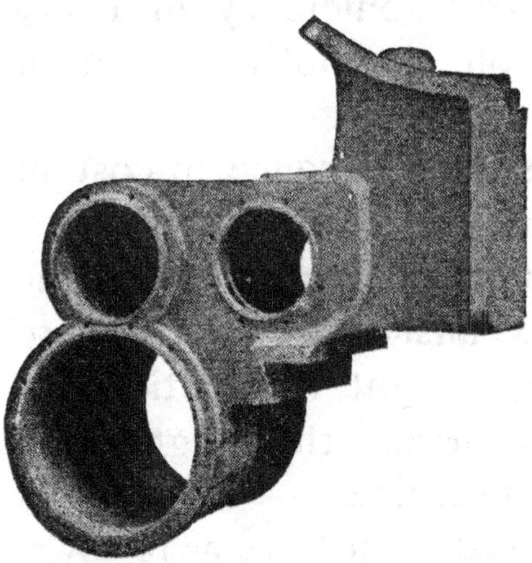

FIGURE 120

FIGURE 121

cylinders is of the piston type, working in a cylindrical steam-chest located in the saddle of the cylinder casting between the cylinders and the smoke-box, and as close to the cylinders as convenience will permit.

As the steam-chest must have the necessary steam passages cast in it and dressed accurately to the required sizes, the main passages in the cylinder casting leading thereto are cast wider than the finished ports. The steam-chest is bored out enough larger than the diameter of the valve to permit the use of a hard cast iron bushing (Fig. 123). This bushing is forced into the steam-chest under such pressure as to prevent the escape of steam from one steam passage to another except by the action of the valve. Thus an opportunity is given to machine accurately all the various ports, so that the admission of steam is uniform under all conditions of service.

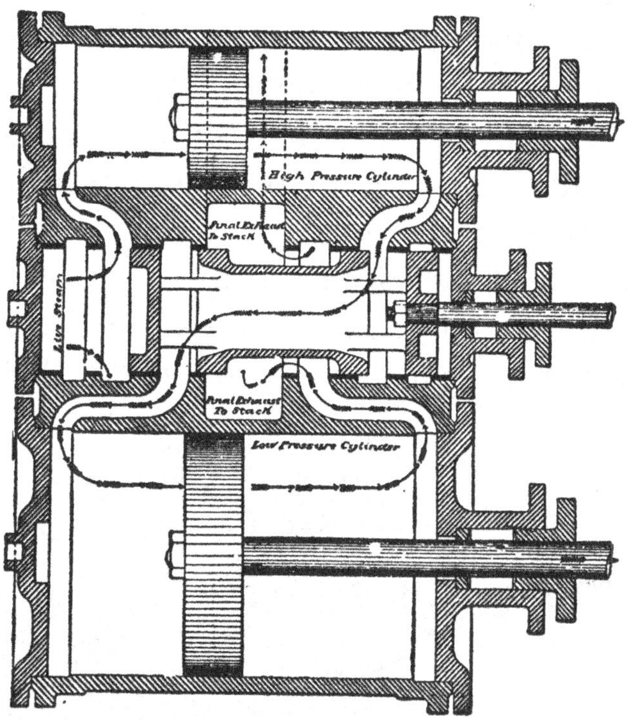

FIGURE 122

The valve, which is of the piston type, double and hollow, as shown in Fig. 124, controls the steam admission and exhaust of both cylinders. The exhaust steam from the high-pressure cylinder becomes the supply steam for the low-pressure cylinder. As the supply steam for the high-pressure cylinder enters

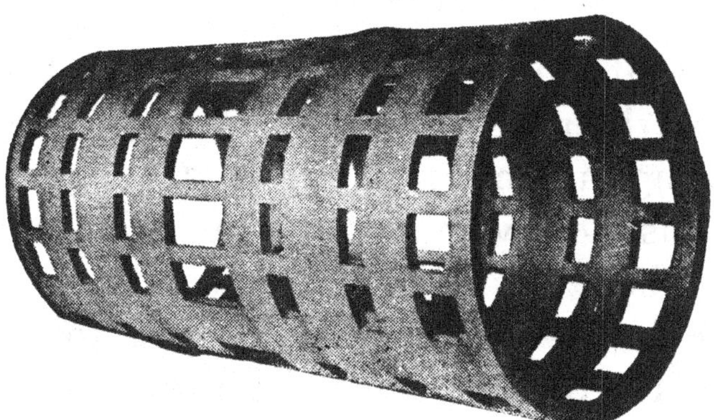

FIGURE 123

the valve-rod should be broken, as it holds them together. Cases are reported where compound locomotives of this system have hauled passenger trains long distances with broken valve-stems and broken valves, the parts being kept in their proper relation while running by the compression due to the variation mentioned. To avoid the possibility of breaking, it is the present practice to pass the valve-stem through the valve and secure it by a nut on the front end.

Cast iron packing rings are fitted to the valve and constitute the edges of the valve. They are pre-

FIGURE 124

vented from entering the steam-ports when the valve is in motion by the narrow bridge across the steam-ports of the bushing, as shown in Fig. 123. The operation of the valve is clearly shown in Fig. 122, the direction of the steam being indicated by arrows.

When the low-pressure cylinder is on top, as shown in. Fig. 121, the double front rail prevents the use of the ordinary rock-shaft and box, and the valve motion is then what is called "direct acting," changing the location of the eccentrics on the axle in relation to the crank-pin. When the low-pressure cylinder is underneath, the rock-shaft is employed, and the eccentrics are placed in the usual position; the valve motion is termed "indirect acting." Fig. 125 shows the relation of the eccentrics with and without the rocker-shaft. Great care should be taken by mechanics, when setting the valves on these locomotives, to observe this difference and not get the eccentrics improperly located on the axle. If the crank-pin is placed on the forward center, the eccentric-rods will not be crossed when the rocker-arm or indirect motion is used, but will be crossed when no rocker-arm or direct motion is used. Serious complications have arisen from this being disregarded.

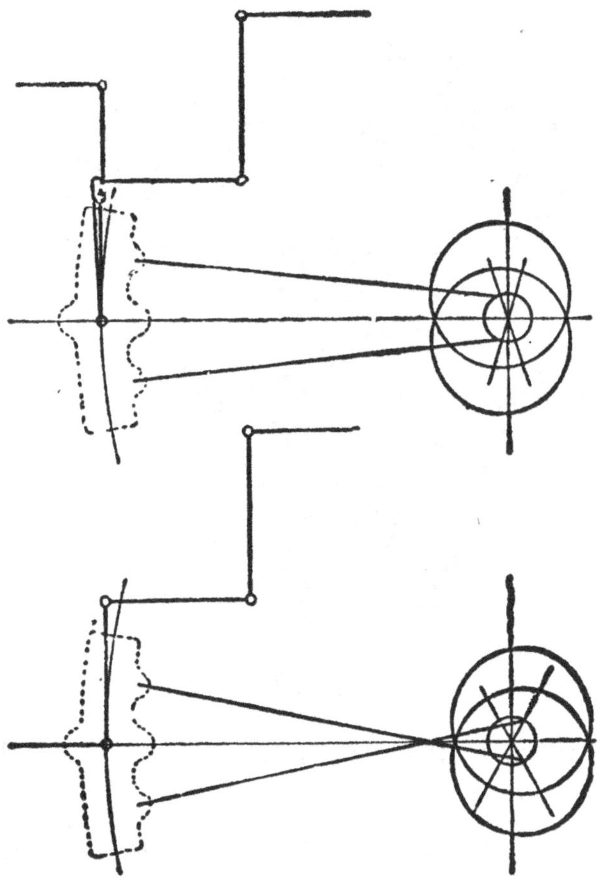

FIGURE 125

In setting the piston valves, only the high-pressure ports are to be considered. Both heads of the steam chest are removed, and with a tram, from some point on the body of the cylinder to the valve stem, the line and line positions of the valve in both front and back motion, are laid off and indicated by a prick punch mark on the valve stem. Using the same tram, the position of the valve at different parts of the stroke can be ascertained, and the opening of the ports noted by the distance from the point of the tram to the prick punch mark. The relation of the low pressure ports to the valve must be ascertained by measurement, the same as the exhaust ports in ordinary slide valves.

FIGURE 126

Various methods have been employed to transfer the motion from the links to the valve-rod. That which has proved most satisfactory is to attach the ends of the link and valve-rods to the arms of an intermediate oscillating shaft. This arrangement allows for the free vertical movement of the end of the rod attached to the link, and gives a parallel movement to the valve-rod. It also makes it convenient to obtain any required lateral variation in the line of the two rods. These parts are thoroughly case-hardened, and with reasonable care should wear indefinitely. It

FIGURE 127

is preferable, however, to use a rock-shaft when possible, as there is then less departure from ordinary locomotive practice.

The cross-head is shown in Fig. 126. It is made of open-hearth cast steel and is machined accurately to

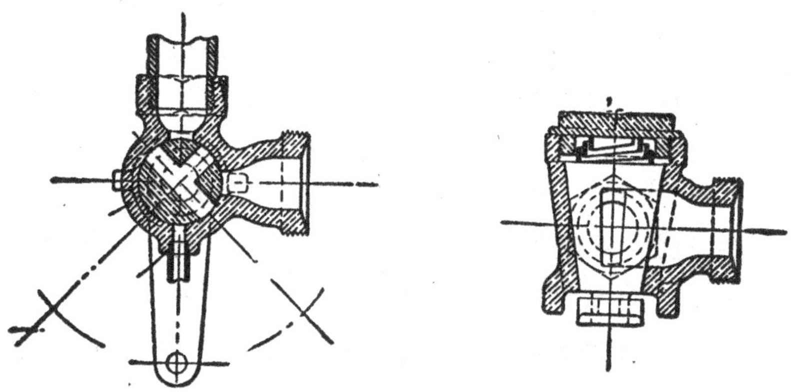

FIGURE 128

size. The bearings for the guide-bars are covered with a thin coating of block tin, about one-sixteenth inch thick, which wears well and prevents heating. The holes for the piston-rods are bored so that the piston-rods will be perfectly parallel, and are tapered to insure a perfect fit.

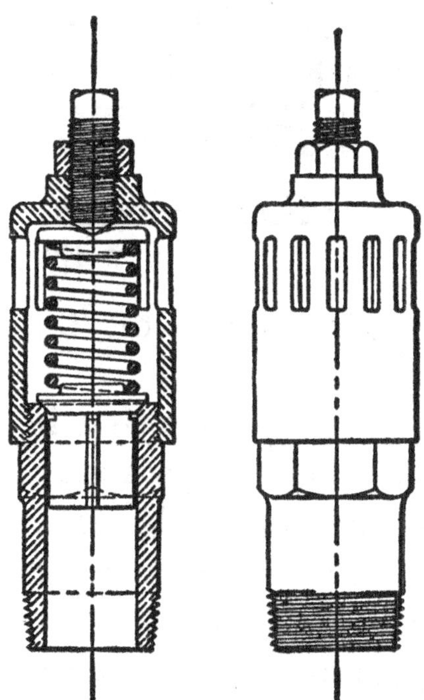

FIGURE 129

The piston shown in Fig. 127 is made with either cast iron or cast steel heads, and is as light as possible. The rods, which are of triple-refined iron, are ground perfectly true to insure good service in connection with metallic packing for the stuffing boxes. The diameter of both piston-rods is the same, both having equal work to perform. They are made large enough to resist strains due to any unequal pressure that may come upon them in starting the locomotive from a state of rest. The cross-head end has a shoulder which prevents the piston-rod being forced into the cross-head, and at the same time permits the cross-head end and the body of the piston-rod to be of one diameter, thus permitting vibratory strains to act throughout the entire length of the rod instead of concentrating them at the shoulder next to the cross-head. The piston-rods are secured to the cross-head by large nuts, and these in turn are prevented from coming loose by taper keys driven tightly against them.

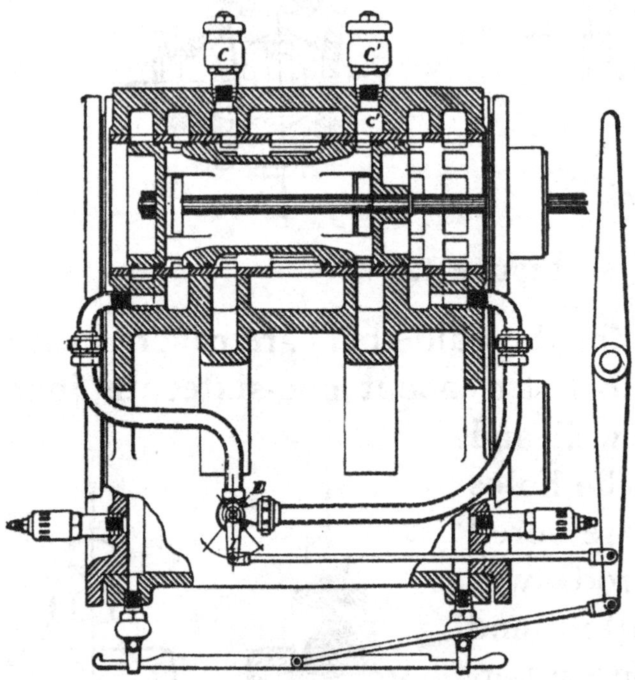

FIGURE 130

It is obvious that in starting these locomotives with full trains from a state of rest, it is necessary to admit steam to the low-pressure cylinder as well as to the high-pressure cylinder, which is accomplished by the use of a starting valve (Fig. 128). This is merely a pass-by valve which is opened to admit steam to pass from one end of the high-pressure cylinder to the other end and thence through the exhaust to the low-pressure cylinder. This is more clearly shown at E in Figs. 130 and 131. The same cock acts as a cylinder cock for the high-pressure cylinder and is operated by the same lever that operates the ordinary cylinder cocks, thus making a simple and efficient device and one that need not become disarranged. This valve should be kept shut as much as possible, as its indiscriminate use reduces the economy and makes the locomotive "logy."

As is usual in all engines, air valves are placed in the main steam passage of the high-pressure cylinder. Additional air valves, marked C and C^1 in Fig. 130, are placed in the steam passages of the low-pressure cylinders to supply them with sufficient air to prevent the formation of a vacuum which would draw cinders into the steam-chest and cylinders.

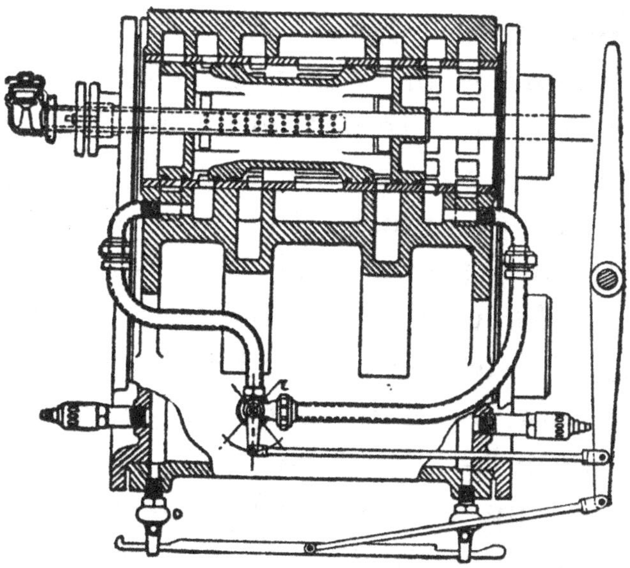

FIGURE 131

The hollow valve stem shown in Fig. 131 accomplishes the same result, but with a more direct action, and is preferable for fast service. The check valve at the end of the hollow stem outside the steam chest is closed by the pressure of the steam, but stands open when the pressure is relieved and air is allowed to pass into the valve through the perforation in the hollow stem. A vacuum is thus prevented from forming in the valve or low-pressure passages.

This arrangement will also prevent the accidental starting of the locomotive occasioned by a leaky throttle. The steam as it slowly escapes will pass through the hollow stem to the open air without creating pressure in the cylinders.

Water relief valves (Fig. 129) are applied to the low-pressure cylinders and attached to the front and back cylinder heads to prevent the rupture of the cylinder in case a careless engineer should permit the cylinders to be charged with water, or to relieve excessive pressure of any kind.

In all other respects the locomotive is the same as the ordinary single-expansion locomotive.

Operation. It is not surprising, in view of their differences of opinion respecting single-expansion locomotives, that there has been much controversy among engineers and firemen in regard to the operation of compound locomotives of this system. The first thing the engineer must learn is to use the reverse lever for what it is intended; that is, he must not hesitate to move it forward when ascending a grade if the locomotive shows signs of slowing up. The reverse quadrant is always so made that it is impossible to cut off steam in the high-pressure cylinder at less than half stroke, which avoids the damage that might ensue from excessive compression. It is perfectly practicable to operate the engine at any position of the reverse lever between half stroke and full stroke, without serious injury to the fire. When starting the locomotive from a state of rest, the engineer should always open the cylinder cocks to relieve the cylinders of condensation, and as the starting valve is attached to the cylinder cocks, this movement also admits steam to the low-pressure cylinder and enables the locomotive

Atlantic Type Locomotive, Baltimore & Ohio Railroad

to start quickly and freely. In case the locomotive is attached to a passenger train and standing in a crowded station, or in some position where it is undesirable to open the cylinder cocks, the engineer should move the cylinder cock lever in position to permit live steam to pass by into the low-pressure cylinder, thus enabling the locomotive to start quickly and uniformly, without any of the jerking motion so common in two-cylinder or cross-compound locomotives. After a few revolutions have been made and the cylinders are free from water caused by condensation or priming, the engineer should move the cylinder cock lever into the central position, causing the engine to work compound entirely. This should be done before the reverse lever is disturbed from its full gear position. The reverse lever should never be "hooked up," thereby shortening the travel of the valve, until after the cylinder cock lever has been placed in the central position. It is often necessary to open the cylinder cocks when at full speed, to allow water to escape from the cylinders, especially when the engineer is what is commonly called a "high-water" man, and in such case no disadvantage is experienced and the reverse lever need not be disturbed. The starting device should not be used for any purpose other than the "starting" of the train. After the train is in motion it should not be used. Cases have been observed where the engineers use it all the time and have the reverse lever "hooked up" in the top notch (half stroke), in consequence of which the locomotive will slow down to a low speed whilst burning an excessive amount of coal. Such running must result in general dissatisfaction.

The starting device is useful in emergencies, as, for

MOGUL FREIGHT LOCOMOTIVE, CHICAGO & ALTON RAILWAY

LOCOMOTIVE EQUIPPED WITH THE YOUNG ROTATIVE VALVE

instance, when stalling with a heavy train on a grade, if live steam is admitted to the low-pressure cylinder sufficient additional power is obtained to start the train and take it over the grade. This should be resorted to only in emergencies, and allowance should be made for the extra repairs caused by frequent cases of this kind.

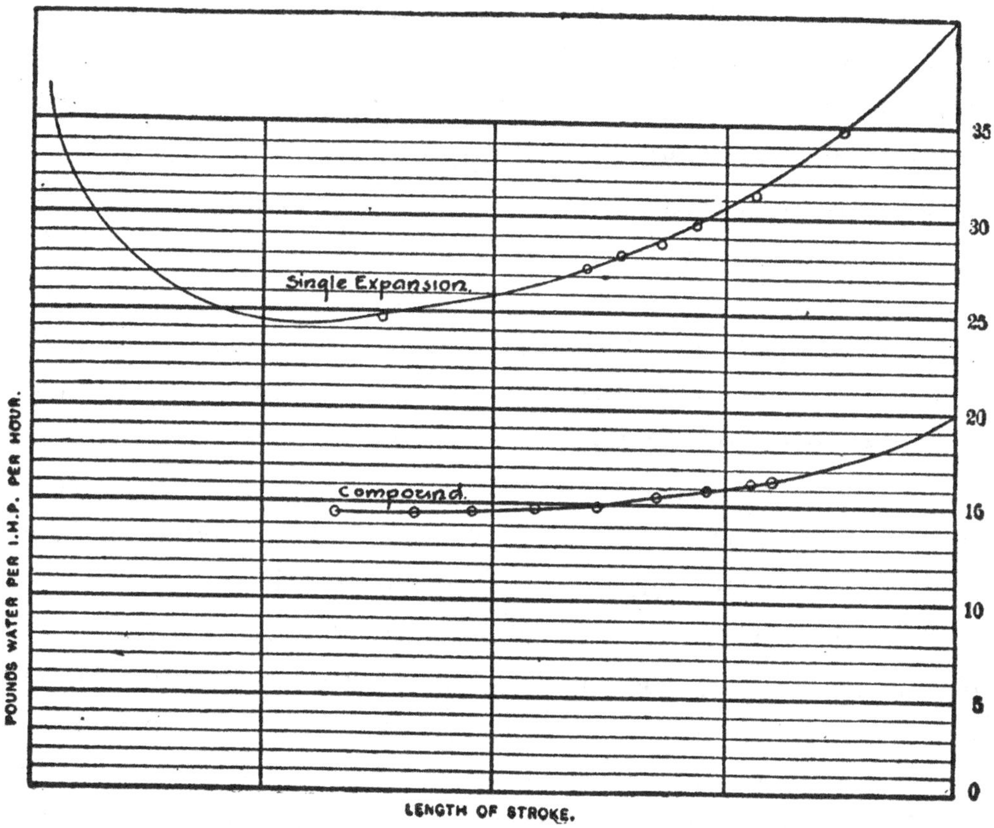

FIGURE 132

On account of the very mild exhaust, the fireman should carry the fire as light as possible. A little practice will enable him to judge how to get along with the least amount of fuel.

The diagram (Fig. 132) shows the difference in the amount of water required to do the work at various points of cut-off in compound and single-expansion locomotives. The upper line shows the rate of water consumption per horse-power developed for several points of cut-off in single-expansion locomotives, whilst the lower line shows the same for compound locomotives. It will be observed that the most economical point of cut-off is about one-quarter stroke on the single-expansion locomotive, and about five-eighths stroke on the compound locomotive. It is also noticeable that the water-rate per horse-power varies very little on the compound locomotive when the reverse lever is moved towards full gear or longer cut-off, but in the single-expansion engine it increases rapidly, causing engineers to remark that they cannot "drop her a notch" on account of "getting away with the water." This does not occur with the compound locomotive when the reverse lever is moved forward towards full gear, and no engineer should open the pass-by valve, admitting live steam to the low-pressure cylinder, until the last notch has been used on the quadrant and the engine is about to stall.

It is also desirable to move the reverse forward a notch before the locomotive slows down too much, as it is better to preserve the momentum of the train than to slow down and again have the trouble of accelerating. In this way both coal and water are wasted. If these instructions are observed the locomotive will work satisfactorily.

Repairs. On account of the great similarity to single-expansion locomotives, mechanics familiar with the latter have no difficulty in understanding these compound locomotives. There is no new element of repairs introduced, no complicated starting or reducing valves, such as are common to other systems of compound locomotives.

The cross-heads, when badly worn, may, in a short

CONSOLIDATION LOCOMOTIVE, LEHIGH VALLEY RAILROAD

time, be retinned by any coppersmith; in fact, an ordinary laborer can be taught this in a few days. The cross-head is heated warm enough to melt solder, and is then cleaned and wiped with solder, using dilute muriatic acid, such as tinsmiths use in soldering. Block tin is then poured against the surfaces so prepared, to which it adheres. A piece of iron placed alongside the cross-head can be used to regulate the thickness

The cross head is then put on a planer to true it up, care being used not to let the tool "dig in" and tear off the tin.

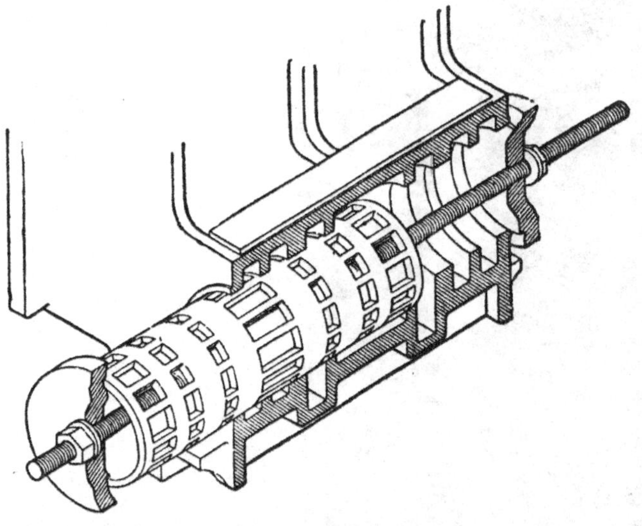

FIGURE 133

The pistons are treated the same as in ordinary single-expansion engines. The packing-rings in the low-pressure cylinder require renewal more frequently than those in high-pressure cylinders. It is also more difficult in compound cylinders to detect faulty packing rings, and they are sometimes noticed only by the locomotive failing in steam and in not making time on the road.

The piston-valves should last a long time if properly lubricated, but when the bushing (Fig. 123) and valve (Fig. 124) are worn enough to require attention, the bushing should be bored out and new rings put in the valve; very often it is not necessary to bore the bushings, merely to put new packing-rings in the valve.

After the bushings (Fig. 123) have been bored several times, larger valves may be fitted to them, so as to have as little play as possible. A very convenient type of boring bar for boring out the bushings has been designed, by which the work can be done without taking down the back head of the steam-chest. It is possible with this tool to bore out the bushings in less time than required to face a valve seat on a single-expansion locomotive.

When putting new bushings in the steam-chests, the device shown in Fig. 133 may be used, which gives the required power and is slow enough to permit the bushing to accommodate itself to the cylinder casting.

When extracting old bushings, it is best to split them with a narrow cape chisel—they are only fit for scrap when removed, and can be much more quickly removed this way than to attempt to draw them out with draw screws.

Enough attention should be given the starting valves to insure their moving in harmony with each other. Engineers sometimes strain the cylinder cock shaft, which causes one starting valve to open and the other to remain shut; this causes the exhaust to beat unevenly, and the engineer is apt to complain that the valves are out of square. Before altering the valve motion on these engines, make sure that the starting valves open and close simultaneously, and examine low-pressure pistons and piston-valve for broken packing-rings. In one case an engineer ran his locomotive two days without any piston-head on one of the low pressure pistons, and even then could not tell what was the matter, only that the locomotive sounded "lame" and did not make good time with the train.

Men were put to work to locate the trouble, and found it, to the great surprise of the engineer.

Suggestions for Running a Vauclain Four-Cylinder Compound Locomotive. In starting the locomotive with a train, place the reverse lever in full forward position, throw the cylinder-cock lever forward, which operation opens the starting-valve and allows live steam to pass to the low-pressure cylinder. The throttle is then opened, and as soon as possible when the cylinders are free of water and the train is under good headway, the cylinder cocks and starting-valve should be closed. As the economy of a compound locomotive depends largely on its greater range of expansion, the engineer should bear in mind that in order to get the best results he must use his reverse lever. After the starting-valve is closed and as the speed of the train increases, the reverse lever should be hooked back a few notches at a time until the full power of the locomotive is developed. If after moving the reverse lever to the last notch, which cuts off the steam at about half stroke in the high-pressure cylinder, it is found that the locomotive develops more power than is required, the throttle must be partially closed and the flow of steam to the cylinder reduced. On slightly descending grades the steam may be throttled very close, allowing just enough in the cylinders to keep the air-valves closed. If the descent is such as to prevent the use of steam, close the throttle and move the reverse lever gradually to the forward notch and move the starting-valve lever to its full backward position. This allows the air to circulate either way through the starting-valve from one side of the piston to the other, relieves the vacuum, and prevents the oil from being blown out of the cylinder. On ascending grades with

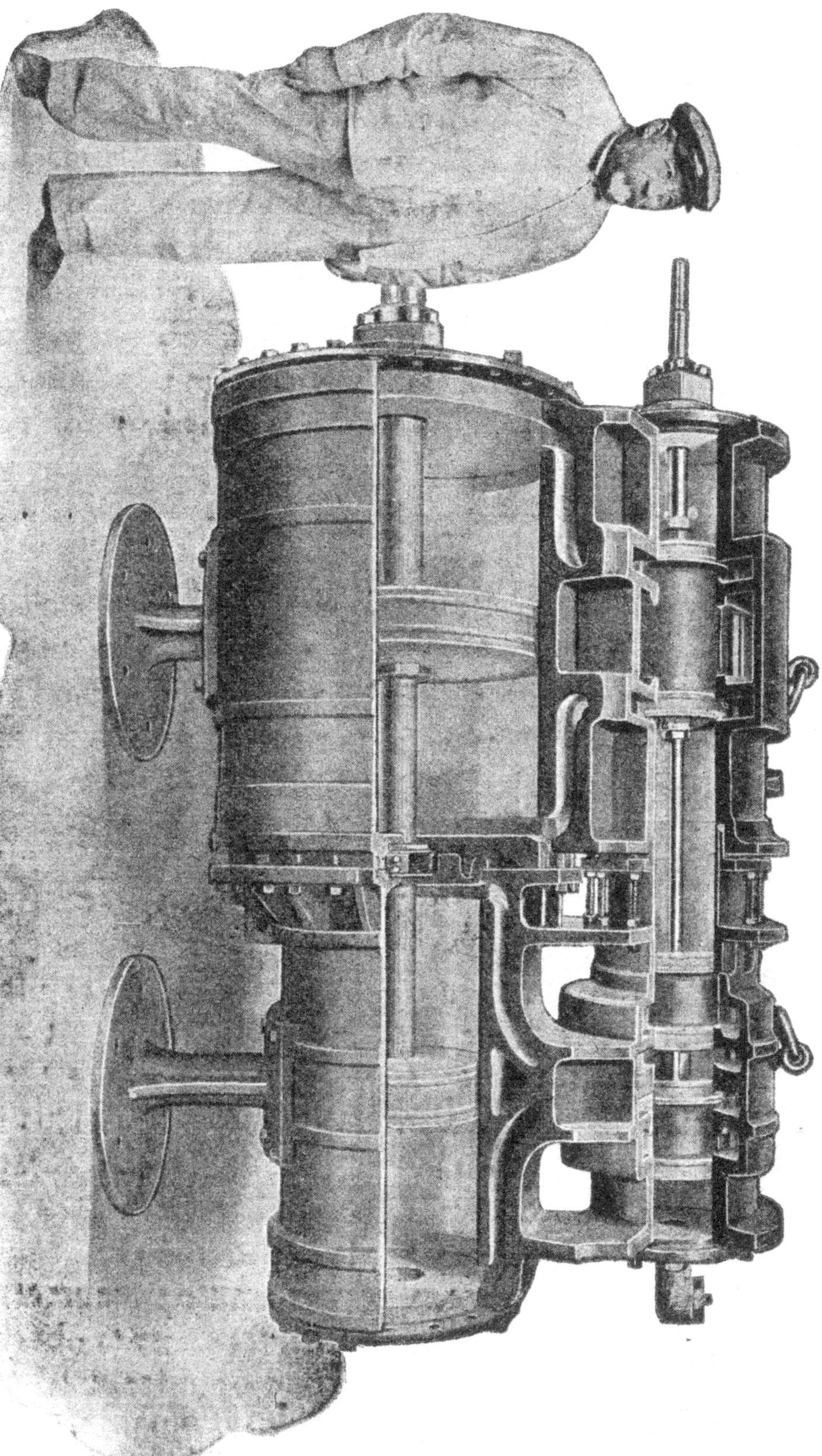

Model of Actual Size, Showing Two Locomotive Cylinders and Valves of the Vauclain Tandem Compound System

heavy loads as the speed decreases the reverse lever should be moved forward sufficiently to keep up the required speed. If, after the reverse lever is placed in the full forward notch, the speed still decreases and there is danger of stalling, the starting-valve may be used, admitting steam to the low-pressure cylinders. This should be done only in cases of emergency and the valve closed as soon as the difficulty is overcome.

The tractive power of Vauclain four-cylinder compound locomotives may be ascertained by the following formula:

$$\frac{C^2 \times S \times \frac{2}{3} P}{D} + \frac{c^2 \times S \times \frac{1}{4} P}{D} = T, \text{ in which}$$

C = Diameter of high-pressure cylinder in inches.

c = Diameter of low-pressure cylinder in inches.

S = Stroke of piston in inches.

P = Boiler pressure in pounds.

D = Diameter of driving wheels in inches.

T = Tractive power.

It is not claimed for compound locomotives that a heavier train can be hauled at a given speed than with a single-expansion locomotive of similar weight and class. No locomotive can haul more than its adhesion will allow; but the compound will, at very slow speed on heavy grades, keep a train moving where a single-expansion locomotive will slip and stall. This is due to the pressure on the crank-pins of the compound being more uniform throughout the stroke than is the case with the single-expansion locomotive.

The principal object in compounding locomotives is to effect fuel economy, and this economy is obtained—

1. By the consumption of a smaller quantity of steam in the cylinders than is necessary for a single-expansion locomotive doing the same work.

2. The amount of water evaporated in doing the same work being less in the compound, a slower rate of combustion combined with a mild exhaust produces a higher efficiency from the coal burned.

In a stationary engine, which does not produce its own steam supply, it is of course proper to measure its efficiency solely by its economical consumption of steam. In an engine of this description the boilers are fired independently, and the draft is formed from causes entirely separate and beyond the control of the escape of steam from the cylinders; hence, any economy shown by the boilers must of necessity be separate and distinct from that which may be effected by the engine itself. In a locomotive, however, the amount of work depends entirely upon the weight on the driving-wheels, the cylinder dimensions being proportioned to this weight; and whether the locomotive is compound or single-expansion, no larger boiler can be provided, after allowing for the wheels, frames, and other mechanism, than this weight permits. Therefore, the heating surfaces and grate area are practically the same in both types, and the evaporative efficiency of both locomotives is determined by the action of the exhaust, which must be of sufficient intensity in both cases to generate the amount of steam necessary for utilizing, to the best advantage, the weight on the driving-wheels. This is a feature that does not appear in a stationary engine, so that the compound locomotive cannot be judged by stationary standards, and the only true comparison to be made is between locomotives of similar construction and weight, equipped in one case with compound and in the other with single-expansion cylinders.

One of the legitimate advantages of the compound

system is that, owing to the better utilization of the steam, less demand is made upon the boiler, which enables sufficient steam-pressure to be maintained with the mild exhaust, due to the low tension of the steam when exhausted from the cylinders. This milder exhaust does not tear the fire, nor carry unconsumed fuel through the flues into the smoke-box and thence out of the smoke-stack, but is sufficient to maintain the necessary rate of combustion in the fire-box with a decreased velocity of the products of combustion through the flues.

The heating surfaces of a boiler absorb heat units from the fire and deliver them to the water at a certain rate. If the rate at which the products of combustion are carried away exceeds the capacity of the heating surfaces to absorb and deliver the heat to the water in the boiler, there is a continual waste that can be overcome only by reducing the velocity of the products of combustion passing through the tubes. This is effected by the compound principle. It gives, therefore, not only the economy due to a smaller consumption of water for the same work, but the additional economy due to slower combustion. It is obvious that these two sources of economy are interdependent.

The improved action of the boiler can be obtained only by the use of the compound principle, while the use of the compound principle enables the locomotive to develop its full efficiency under conditions which in a single-expansion locomotive would require a boiler of capacity so large as to be out of the question under the circumstances usually governing locomotive construction. It is therefore evident that where both locomotives are exact duplicates in all their parts, excepting the cylinders, the improved action of the boiler is due entirely to the compound principle, and the percentage of economy should be based upon the total saving in fuel consumption, and not upon the water consumption, as in stationary practice.

For the benefit of those who may test these locomotives, the following method is presented of determining the water rate per horse-power from an indicator diagram:

S = Stroke in inches.
C = Per cent of stroke completed at cut-off.
P = Pressure of steam at cut-off, taken from zero.
Wp = Weight per cubic foot of steam at P pressure.
H = Per cent of stroke uncompleted at compression.
Q = Pressure of steam at compression, taken from zero.
Wq = Weight per cubic foot of steam at Q pressure.
E = Per cent of clearance in H.-P. cylinders.
A = Area of H.-P. cylinders.
P = M.E.P. of H.-P. cylinders.
a = Area of L.-P. cylinders.
K = M.E.P. of L.-P. cylinders.
N = Number of revolutions per minute.

r = Ratio $\frac{a}{A}$; hence, $a = A \times r$.

All calculations are made on the basis of the high-pressure cylinder doing the work of both cylinders.

The volume of the piston displacement is AS, and the volume at cut-off is $A\,S\,C$, since C is the proportion of stroke completed at cut-off. The volume of N revolutions would be $A\,N\,S\,C$. As there are two strokes of the piston for each revolution, and there is an engine on each side of the locomotive, assuming that both engines are doing exactly the same work, there would be four strokes per revolution; hence $4\,A\,N\,S\,C$ is the volume of piston displacement at cut-off for one revolution. Since the clearance-space is expressed in percentage of the piston displacement of one stroke, and this space is filled at each stroke, the volume of the clearance-space for one revolution would be $4\,ANSE$. The sum of these two quantities divided by 1728 will give the volume in cubic feet. The indicator-card gives the pressure at cut-off, and a reference to Table 4 will give the weight of steam at that pressure; hence, the amount of steam used per revolution becomes $\left(\frac{4ANSC+4ANSE}{1728}\right) Wp$. But there is a certain amount of steam saved at compression, and the volume at this point would be $\left(\frac{4ANSH+4ANSE}{1728}\right) Wq$, the volume of the clearance space being again taken into consideration. Since this steam is saved by compression, it should be deducted from the amount used, and the formula becomes:

$$\left(\frac{4ANSC+4ANSE}{1728}\right) Wp-\left(\frac{4ANSH+4ANSE}{1728}\right) Wq;$$

or $$\frac{4ANS}{1728}\Big((C+E)\ Wp-(H+E)\ Wq \Big).$$

The H.-P. equals $\frac{4ANS(P+rK)}{12 \times 33{,}000}$.

Then the water rate per minute would be

$$\frac{\frac{4ANS}{1728}\Big((C+E)\ Wp-(H+E)\ Wq\Big)}{\frac{4ANS(P+rK)}{12 \times 33{,}000}},$$

or $$\frac{229.16}{P+rK}\Big((C+E)\ Wp-(H+E)\ Wq \Big);$$

and the rate per hour would be $\frac{60 \times 229.16}{P+rK}$,

or $\frac{13750}{P+rK}\Big((C+E)\ Wp-(H+E)\ Wq \Big)$, which formula is to be used:

If it is desired to get the steam at release H.-P., substitute the value of the point *R* and pressure *t*, also $S \times R$, respectively, for *C*, *p*, and $C \times S$. See Figs. 134 and 134½.

M.E.P. H.-P. cylinder....87 pounds Clearance.. .08
M.E.P. L.-P. cylinder ....32 pounds Ratio..2.87 to 1
M.E P. referred to H.-P. cylinder................178.84
M.E.P. referred to L.-P. cylinder.................62.31

$$178.84 = P + rK$$

$$62.31 = K + \frac{P}{r}$$

135.3
14.7
150.0 = .3376 pound per cubic foot of steam at cut-off H.-P. cylinder.

60.3
14.7
75.0 = .1756 pound per cubic foot of steam at compression H.-P. cylinder.

30.
14.7
44.7 = .1079 pound per cubic foot of steam at point on L.-P. expansion line.

16.
14.7
30.7 = .0758 pound per cubic foot of steam at compression L.-P. cylinder.

$$\frac{13750}{178.84} = 76.88 \qquad \frac{13750}{62.31} = 220.67$$

$(.677 + .08) \times .3376 = .2556 \qquad (.238 + .08) \times .1756 = .0558$

.2556
.0558
.1998

$.1998 \times 76.88 = 15.36$ pounds steam at cut-off H.P. cylinder.

$(.744 + .08) \times .1079 = .0889 \qquad (.083 + .08) \times .0758 = .012$

.0889
.0124
.0765

$.0765 \times 220.67 = 16.89$ pounds steam at point on expansion line L.-P. cylinder.

Balanced Compounds. The ideal reciprocating steam engine, stationary or locomotive, simple or compound, is an engine in which the reciprocating parts are perfectly balanced against each other, and that balancing should be accomplished without the aid of rotative counter weights. This can be done only by a correct distribution of the steam to the two or more cylinders, and then transmitting the energy developed in each cylinder, directly through the medium of its own piston rod and connecting rod to the engine shaft. The proper balancing of the reciprocating parts of locomotives has always been an especially serious problem, and has grown more serious with the gradual increase in the size and speed of engines. But American locomotive builders have not been timid in meeting and solving this problem, and to-day the four-cylinder balanced compound locomotive stands forth as a splendid specimen of mechanical ingenuity and skill in designing, and will in time, if given a square deal, prove to be the ideal locomotive.

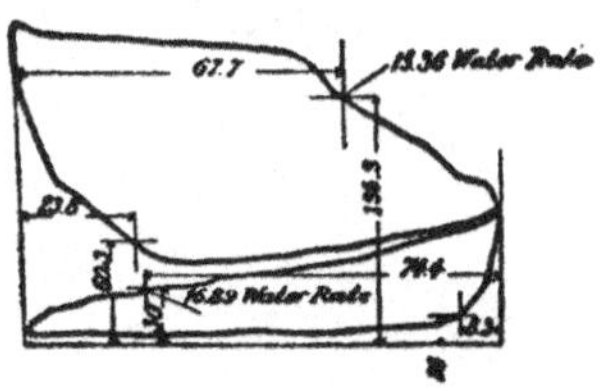

FIGURE 134

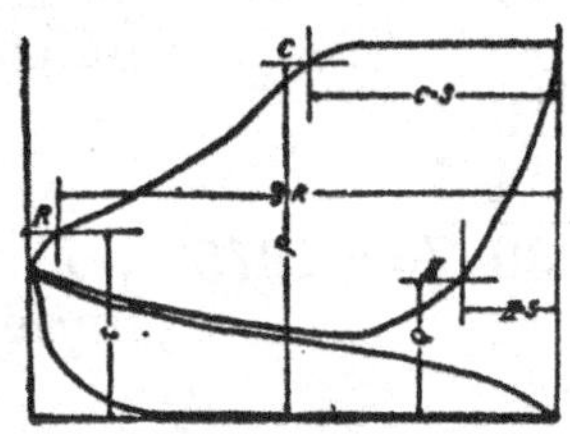

FIGURE 134 a

The Baldwin Locomotive Works kindly supply the following brief description of the four-cylinder balanced compound built by them.

The cylinders are a development of the original Vauclain four-cylinder compound type, with one piston slide valve common to each pair.

Instead of being superimposed and located outside of the frames, the cylinders are placed horizontally in

TEN WHEEL PASSENGER ENGINE, C. R. I. AND P.

line with each other, the low-pressure outside, and the high-pressure inside the frames.

The slide valves are of the piston type, placed above

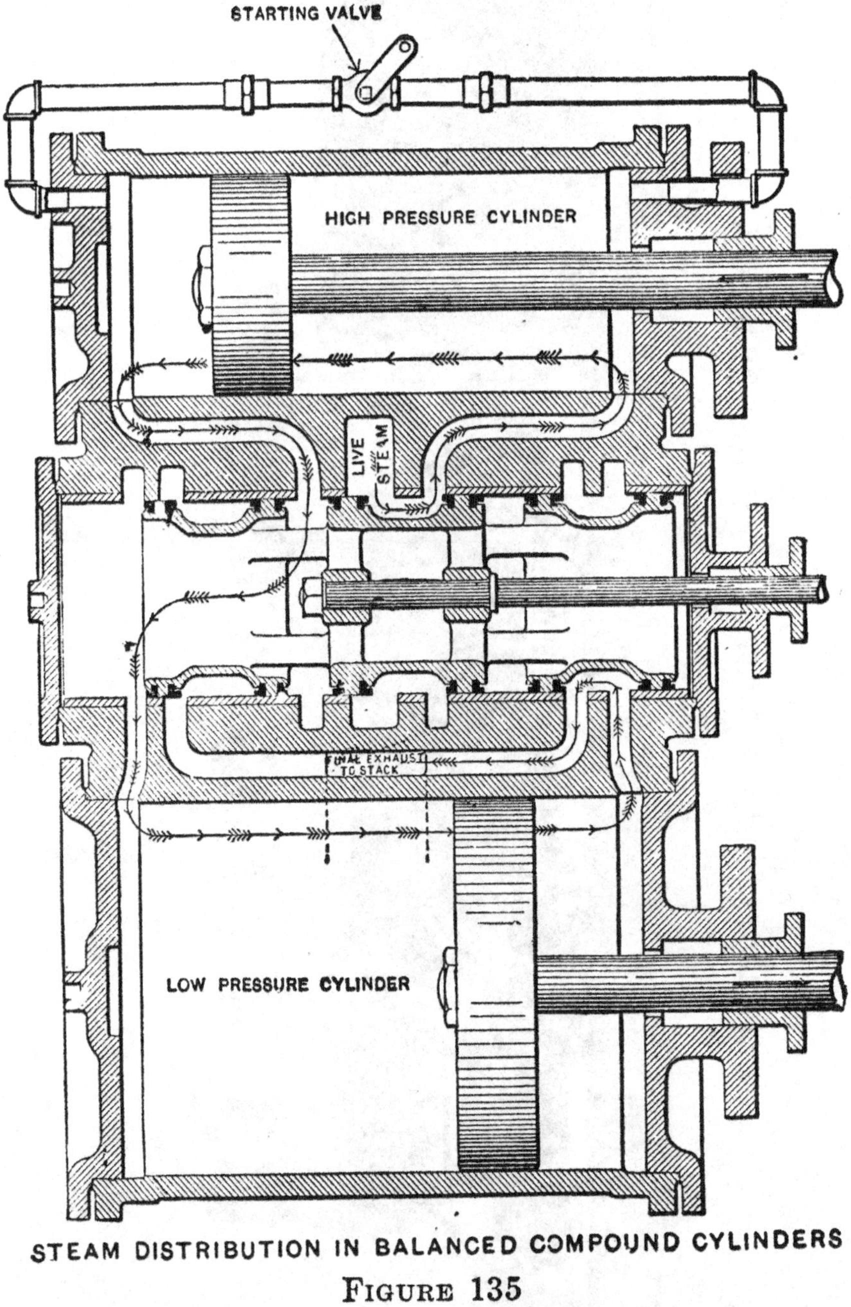

STEAM DISTRIBUTION IN BALANCED COMPOUND CYLINDERS

FIGURE 135

and between the two cylinders which they are arranged to control. A separate set of guides and connections is required for each cylinder.

4 CYLINDER BALANCED COMPOUND,
BUILT FOR N. Y. C. AND H. R. R.

The two high-pressure cylinders being placed inside the frames, the pistons are necessarily coupled to a crank axle. The low-pressure pistons are coupled to crank-pins on the outside of the driving wheels. The cranks on the axle are set at 90° with each other, and at 180° with the corresponding crank-pins in the wheels. The pistons therefore travel in the opposite direction, and the reciprocating parts act against, and balance each other to the extent of their corresponding weight. The distribution of steam is shown in the accompanying diagram (Fig. 135). The live steam port in this design is centrally located between the induction ports of the high-pressure cylinder. Steam enters the high-pressure cylinder through the steam port and the central external cavity in the valve. The exhaust from the high-pressure cylinders takes place through the opposite steam port to the interior of the valve, which acts as a receiver. The outer edges of the valve control the admission of steam to the low-pressure cylinder. The steam passes from the front of the high-pressure cylinder through the valve to the front of the low-pressure cylinder, or from the back of the high-pressure to the back of the low-pressure cylinder. The exhaust from the low-pressure cylinder takes place through external cavities under the front and back portion of the valve, which communicate with the final exhaust port. The starting valve connects the two live steam ports of the high-pressure cylinder to allow the steam to pass over the piston.

The American Locomotive Company build a four-cylinder balanced engine, having the cylinders located in practically the same manner as the Baldwin engine just described.

The use of four cylinders, two high-pressure and two

low-pressure, gives an opportunity for compounding under the most favorable conditions, and with each high-pressure piston working 180° from its low-pressure piston, and the other pair working 90° from the first pair, the successive impulses from the four cylinders produce a remarkably uniform turning moment. This results in a much more rapid rate of acceleration when starting up than has been possible with two-cylinder engines or with many previous types of four-cylinder engines.

The following advantages are claimed for the balanced type of locomotives, by their builders, and the claim appears to be well founded.

1. The elimination of counterbalance weights from the driving wheels, the engine nevertheless being in perfect balance both horizontally and vertically. This results in the complete absence of slip at high speed.

2. The more perfect compounding which results from this arrangement of cylinders, whereby it becomes possible to secure more favorable cylinder volume ratios than with the two-cylinder compound.

3. The consequent approximately uniform turning moment throughout each revolution.

4. The power of quick acceleration, resulting partly from the uniform turning moment and partly from admitting to the low-pressure cylinders, at the time of starting and through a special starting valve, live steam at reduced pressure.

5. The reduction of stresses in the driving axles, crank-pins and other parts of machinery due to the system of distributing power from the cylinders, approximately one-half being transmitted to the forward driving axle and one-half to the rear axle.

6. Increased hauling capacity and endurance at high

speed, due principally to the perfection of the compounding and the consequent economical use of steam, but partly also on account of the perfect balance of the reciprocating and revolving parts.

Tandem Compounds. Theoretically the tandem compound with its four cylinders would appear at first glance to be the ideal design, especially for locomotives, as it places the cylinder in line, and as a result of this the strains are all brought to bear along the same axis. One connecting rod, one set of guides, and one piston rod, serve to reduce the number of parts and, although there are two valves, one for the high-pressure cylinder and one for the low, yet one valve rod operates both valves. Notwithstanding that the tandem compound has all of these and numerous other points in its favor, it does not appear to have grown in popularity in the same degree as have the other types of compound locomotives.

One of the main objections to the tandem, and no doubt a well-founded one, is based upon the difficulties that are encountered in the examination and repair of the pistons and valves.

In many of the designs the methods that must of necessity be employed to do this work are very complicated, and consume too much time to meet with the approval of the "Boss"; and when it comes to running the engine out on the road, there are many engineers who are not studious enough, by nature, to make a success of running a compound, especially of the tandem type. A compound engine, whether marine, stationary, or locomotive, requires careful handling, more so in fact than does a simple engine, and if the engineer in charge of one expects to get good results from her, it is absolutely necessary that he should have at least an elementary knowledge of the principles upon which it is constructed, the routes of the steam passages, the construction of the valves, pistons, etc. This knowledge is easily within the grasp of every engineer, and every fireman who expects to become an engineer, and the opportunities for obtaining it are many.

The tandem compound built by the American Locomotive Company has been quite largely used in freight service during the last five years, and has met with a fair degree of success.

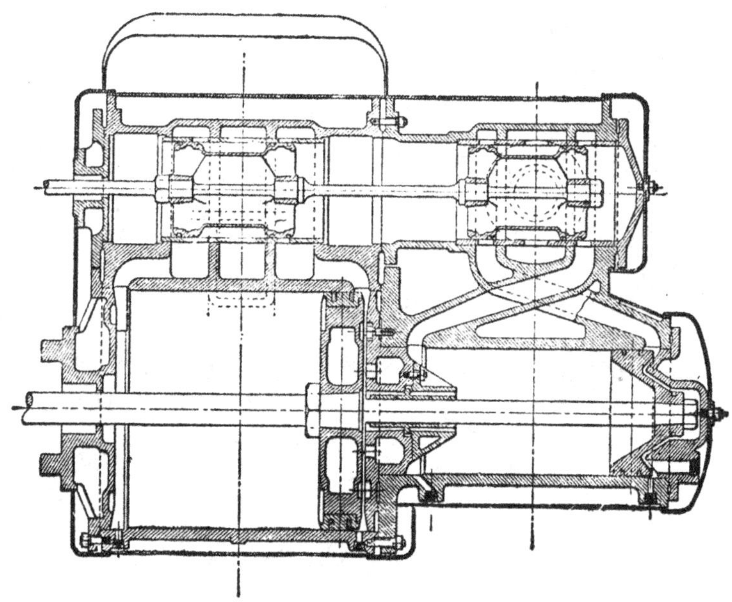

FIGURE 136

Cylinders. The general arrangement of cylinders and of pistons and valves is shown in Fig. 136, in which the high-pressure cylinder is forward of the low-pressure cylinder, with both pistons on the same rod. The steam chest is common to both high and low-pressure cylinders, being open from end to end and serving the purpose of a receiver.

The valves are hollow and permit an unrestricted flow of steam through the steam chest. There being no receiver pipe on these engines, the smoke-box is fitted up with steam pipes and exhaust pipe exactly the same as in simple engines.

Piston Valves. On the high-pressure cylinders the valves are arranged for internal admission, and on the low-pressure cylinders for external admission. An examination of Fig. 136 will show that this design of valves allows steam to be admitted to the same side of each piston by means of the crossed ports on the high-pressure cylinder, the valves being shown as admitting steam.

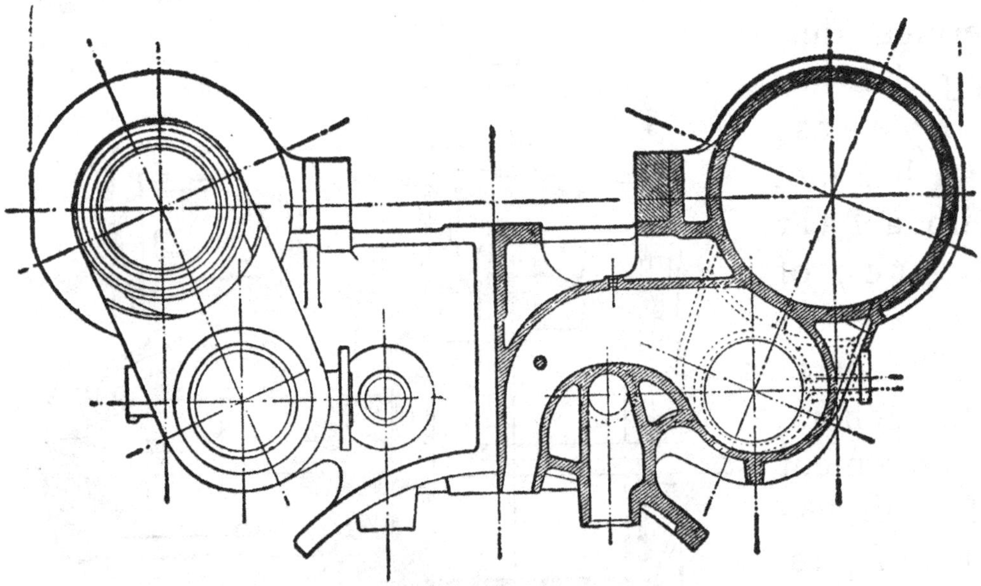

FIGURE 137

Low-Pressure Cylinders. The saddle and cylinders are shown in Fig. 137 in front view and vertical section, in which the coring is shown for steam and exhaust passages. The saddle has an opening cored into the steam-pipe passage, extending from front to back on each side, where there is a circular flange for connection to the short length of steam pipe which extends from front of saddle to the high-pressure cylinder. Coring this passage through from end to end of saddle makes the cylinders interchangeable for use on either side.

Starting Valve. To work the engine, simple or compound, at will, the starting valve shown in Fig. 138 is used, this valve being secured to the side of steam chest over the high-pressure cylinder, and having direct communication with the steam passages into that cylinder. The by-pass valves for the high-pressure cylinders are also contained in the casing of this starting valve and are worked in connection with the latter.

By-Pass Valves. For the purpose of relieving the low-pressure cylinder of excessive pressure when

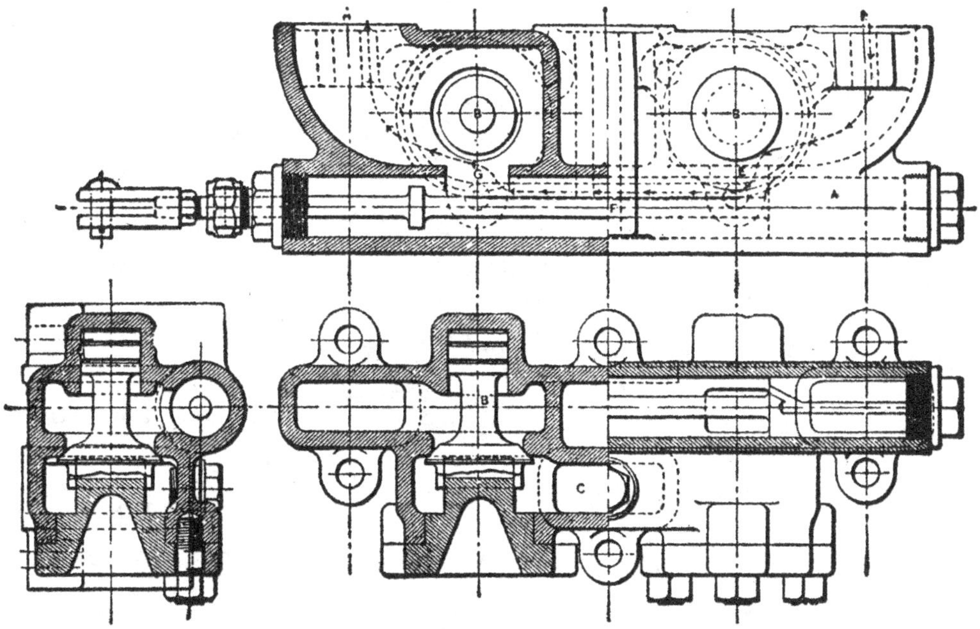

FIGURE 138

working steam, or freeing the same cylinder from back pressure when drifting, the by-pass valves shown in Fig. 139 are used. These by-pass valves are bolted to the side of the steam chest near each end of low-pressure cylinder, and furnish communication between the steam chest and steam ports in cylinder.

Operation, Working Simple. To start the locomotive simple—that is, to admit live steam directly to the low-pressure cylinders—the starting valve A is placed in position shown in Fig. 138 by means of a lever in the cab. Steam is admitted to high-pressure steam chest through the short steam pipe connecting saddle and chest, and passes through ports D and H. which register with the high-pressure steam ports in steam chest. From D the steam is admitted to ports E and G, and passes around the by-pass valves B, B, into port H, the valves B, B, being held up to their seats by pressure from below through port C, which opens directly into the steam chamber of chest. Steam, having access to both high-pressure steam ports, passes through both hollow piston valves and is admitted to the low-pressure cylinder, the engine working as a simple locomotive.

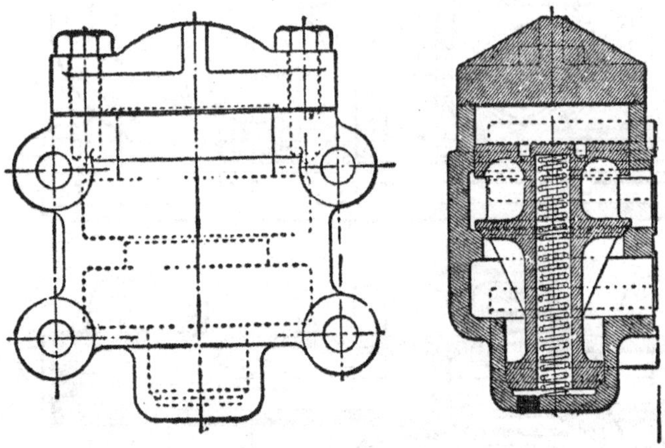

FIGURE 139

Working Compound. When working compound, the starting valve A in Fig. 138 is brought to lap on port E, shutting off high-pressure steam from its passage into the low-pressure end of steam chest. Under these conditions no steam can reach the low-pressure cylinder, except from the exhaust of the high-pressure cylinder.

Drifting. When drifting or not working steam, the by-pass valves B, B, in Fig. 138, being in a vertical position, fall away from their seats by gravity and give a clear opening between the two ends of the high-pressure cylinder. The by-pass valves in Fig. 139 for the low-pressure cylinders are also in a vertical position, and are held to their seats by the steam chest pressure when working steam. When running with closed throttle, the by-pass valves (Fig. 139) are raised from their seats by any pressure on the lower side, assisted by the spring under valve. With the valves raised from their seats there is a continuous opening between the two ends of low-pressure cylinder through cylinder steam ports into steam chest, providing relief from back pressure when drifting, by equalizing the pressure in the cylinders.

Starting. Any compound engine will do more economical and satisfactory work operated as a compound, and should therefore never be worked as a simple engine except in starting, or when likely to stall on grades, and then only long enough to overcome the resistance of the train.

Water. Attention should be given to the quantity of water carried in the boiler, with the view of using steam as dry as possible. Water should not be any higher over crown sheet than is necessary for safety, since high water is not conducive to economy in operation, and is also a menace to proper lubrication.

Lubrication. When running under steam the high-pressure cylinder should receive the greater amount of oil. When drifting the reverse should be the rule, the low-pressure cylinder having the more oil.

Breakdowns. When necessary to disconnect the engine on the road, the same methods may be used as with a simple engine, as to removal of parts, blocking of crosshead, etc.

Testing Tandem Compound. The illustrations show sections through steam chests, valves and cylinders, with valves in various positions for testing. (Rules were formulated by E. P. Roesch, master mechanic, Chicago & Alton Railroad.)

It will be noticed that high-pressure valve A is central or internal admission, while low-pressure valve B is external or end admission. Also notice that ports C and D, leading from high-pressure steam chest E to cylinder F, are crossed. Both valves A and B, and cylinder packings and piston-packing sleeve G, can be tested on each side of engine by simply moving reverse lever. To make tests, place the engine on quarter on side to be tested and proceed in manner designated on following pages.

Testing High-Pressure Valve. Engine on top quarter. Reverse lever in center of quadrant. Starting valve S closed as in Fig. 146. This places both valves A and B in central position, covering all ports on side to be tested.

By opening throttle, steam is admitted to the high-pressure steam chest E, as shown in shade. If steam now flows from either cylinder cock, H or I, the high-pressure valve A is blowing.

Testing Low-Pressure Valve. Engine on top quarter. Reverse lever on center, as in Fig. 140. Starting valve S open, as in Fig. 145.

Remove by-pass valve M in Fig. 145, but replace valve-cap, which is not shown, as it is bolted to under side of starting valve. This allows steam to flow through by-pass from high-pressure steam chest E, through starting valve ports N and O, and past exhaust edges X and Y of high-pressure valve A, into low-pressure steam chest P.

If steam now blows from both low-pressure cylinder cocks K and L, the low-pressure valve B is leaking.

Testing High-Pressure Cylinder Packing. Engine on

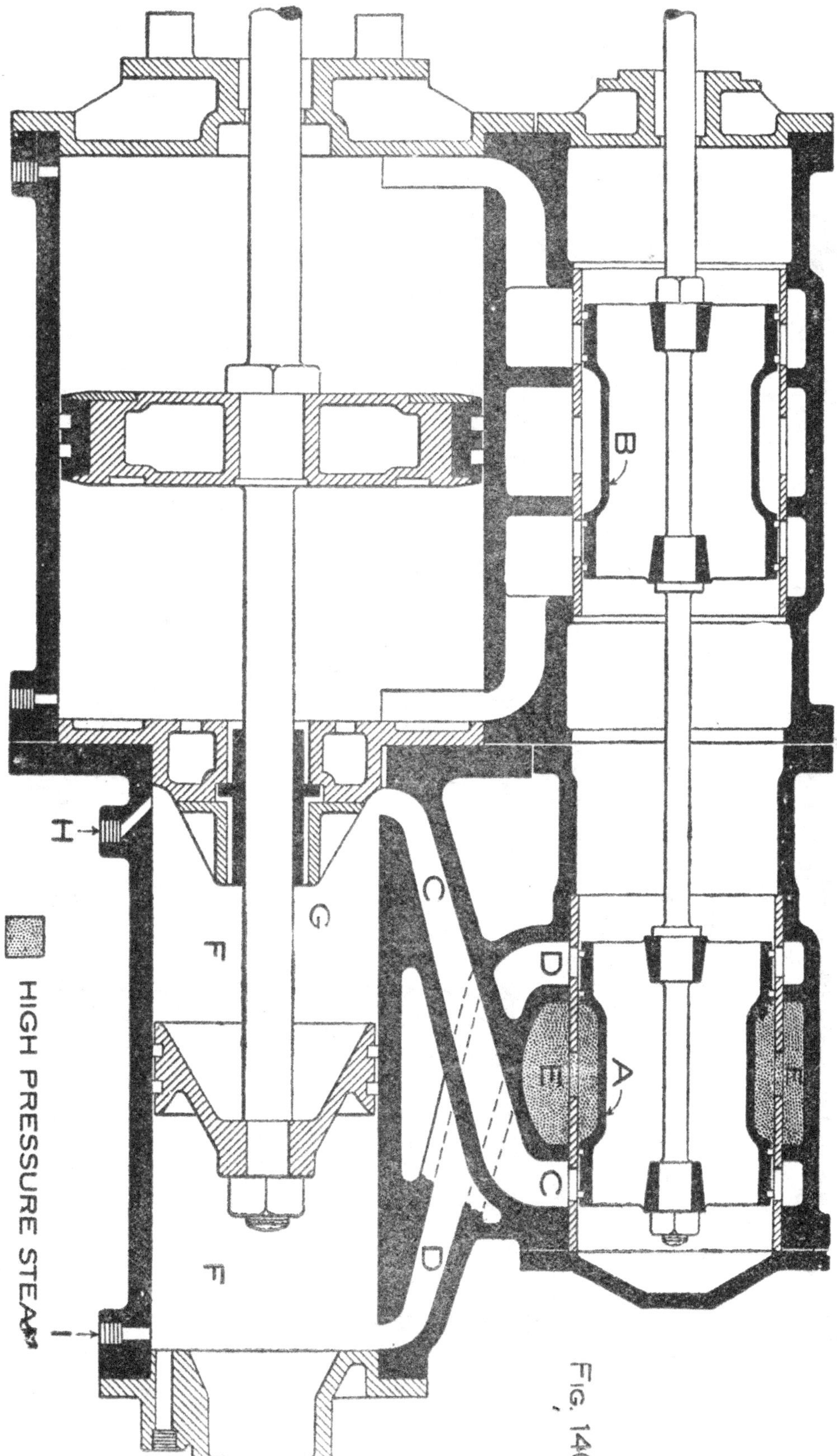

FIG. 140

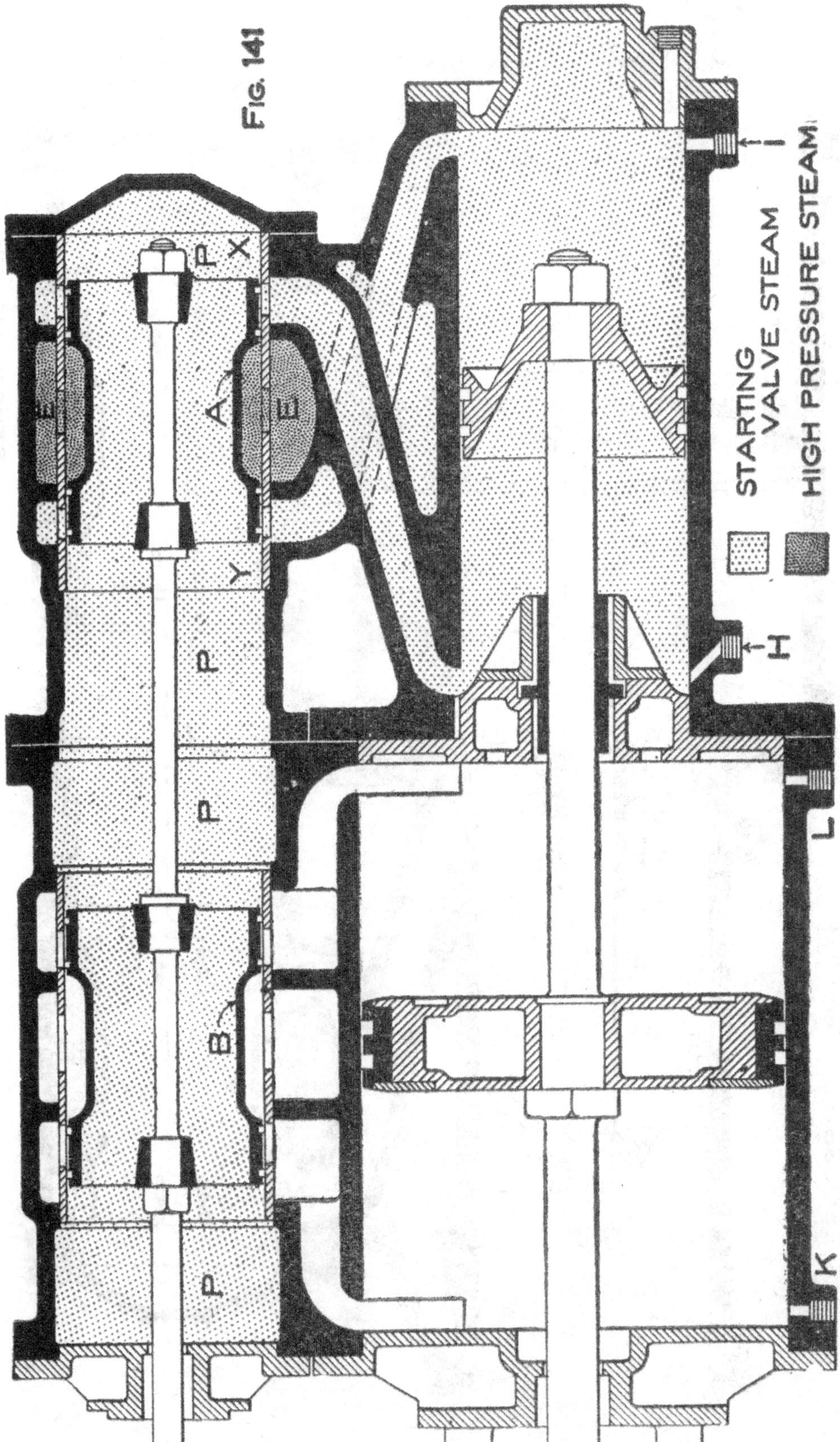

Fig. 141

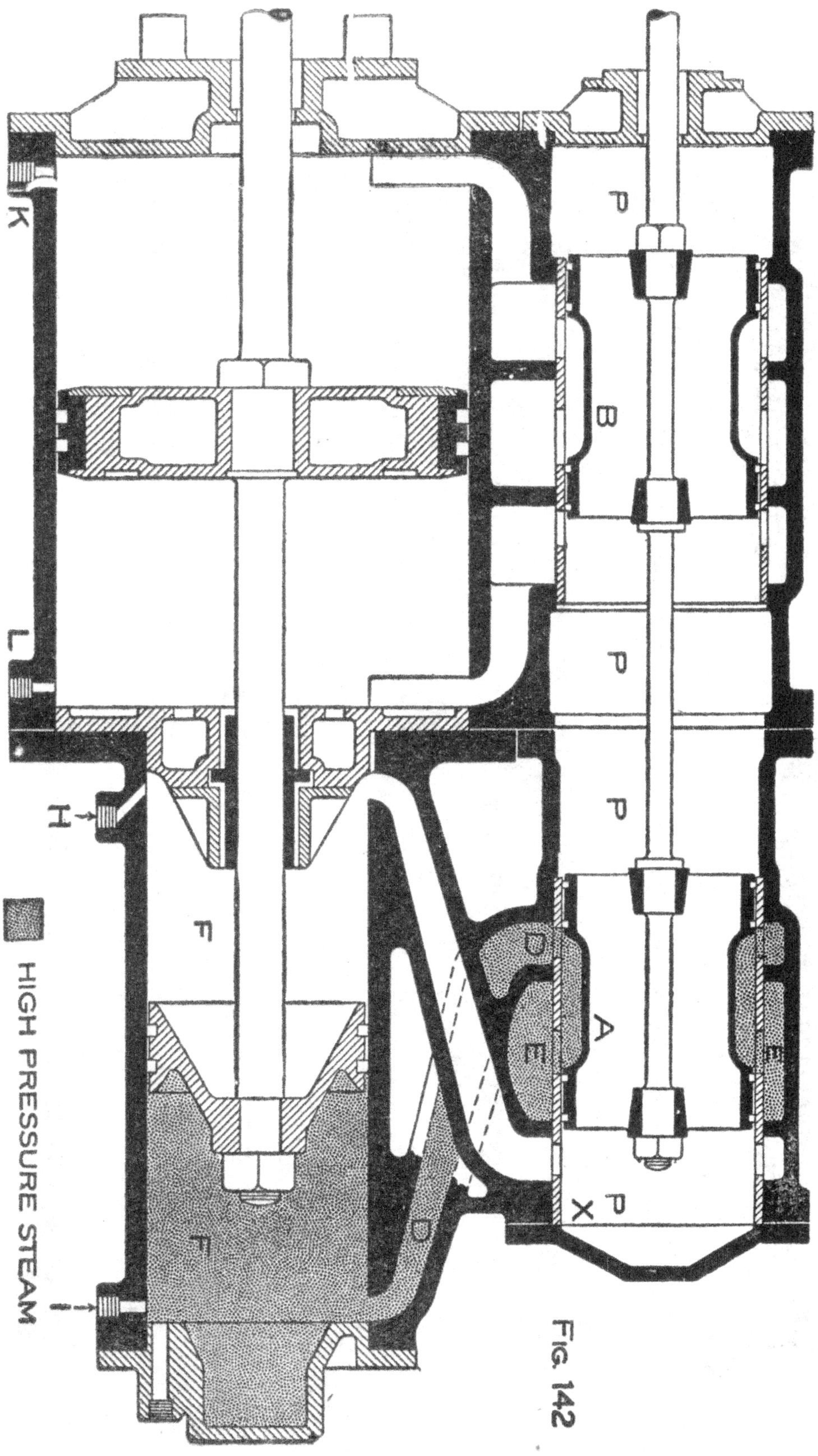

Fig. 142

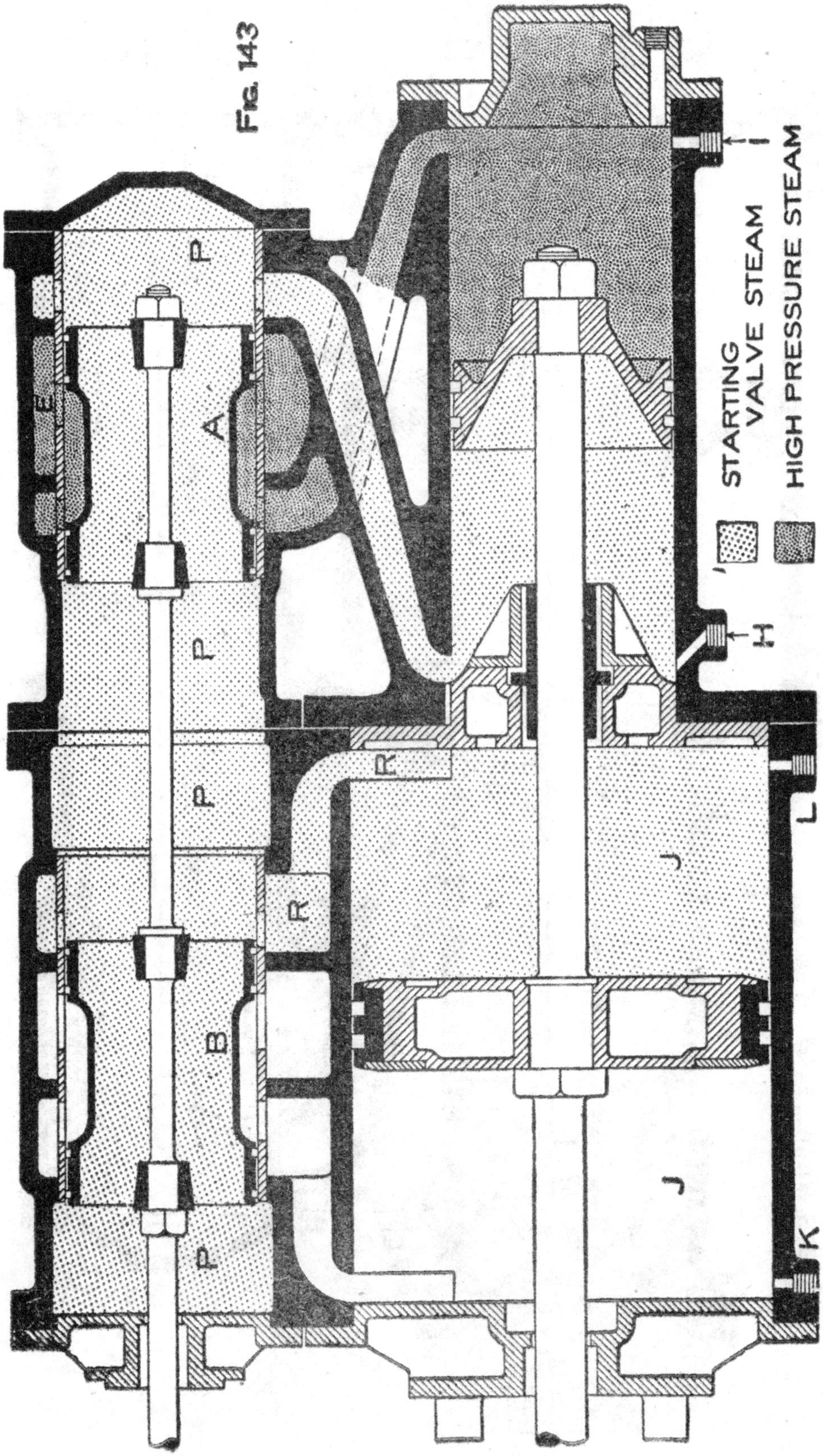

Fig. 143

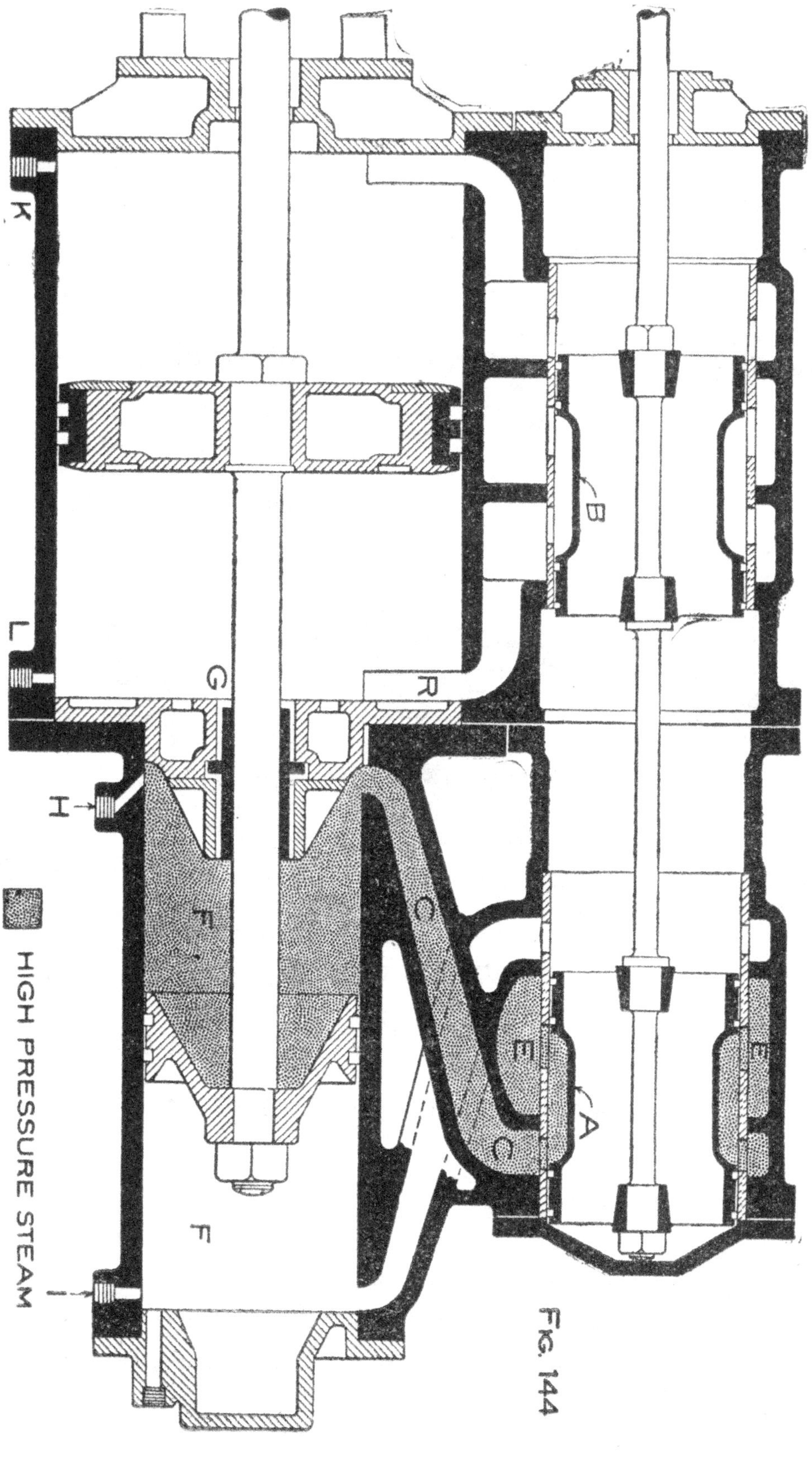

FIG. 144

top quarter. Starting valve S closed, as in Fig. 146. Reverse lever in back motion.

This admits steam from high-pressure steam chest E through steam port D, to front end of high-pressure cylinder F.

If steam now blows from back high-pressure cylinder cock H, the high-pressure piston packing is blowing.

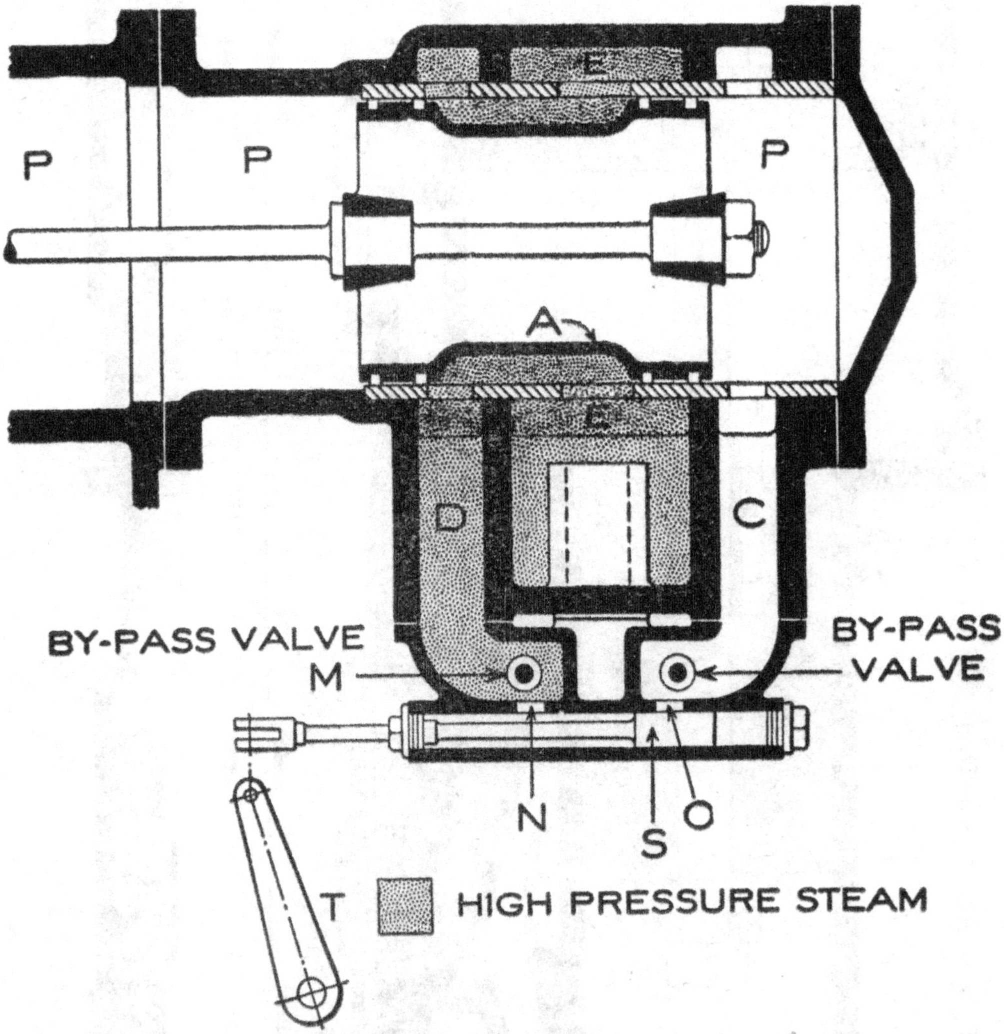

FIGURE 145

Testing Low-Pressure Cylinder Packing. Engine on top quarter. Starting valve S open, as in Fig. 145. Reverse lever in back motion. This allows steam to flow through starting valve into low-pressure steam chest P, thence through front low-pressure steam port R to front end of low-pressure cylinder J.

If any steam shows at back low-pressure cylinder cock K, the low-pressure piston packing is blowing. Always test low-pressure piston packing in this position.

Testing Piston Packing Sleeve, Between Cylinders.

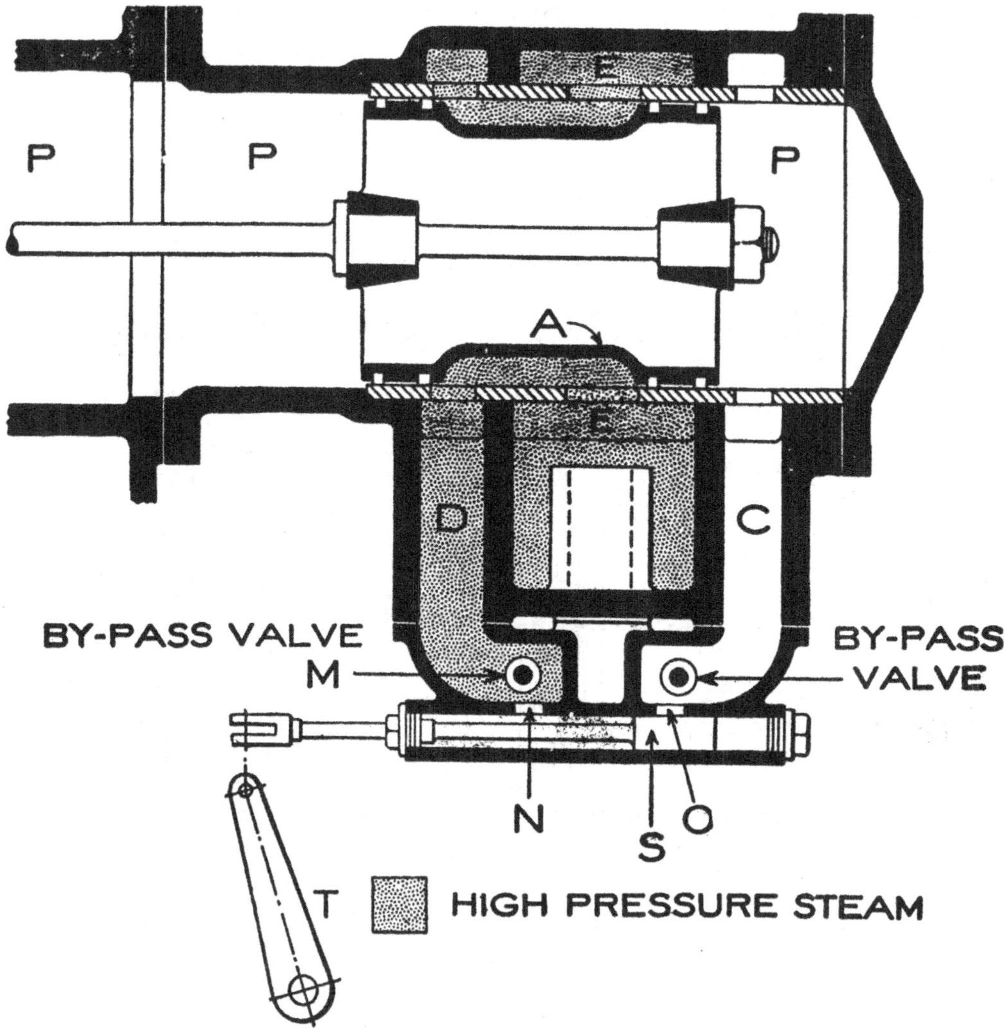

FIGURE 146

Engine on top quarter. Starting valve S closed, as in Fig. 146. Reverse lever in forward motion. This admits steam from high-pressure steam chest E, through steam port C, to back end of high-pressure cylinder F only.

If steam now flows from front low-pressure cylinder cock L, the piston sleeve G is worn and leaking.

Starting Valve in Position for Working Simple. Fig. 145 shows section through high-pressure valve, steam chest and starting valve. By-pass valve M removed, but having valve-cap replaced. For working simple, starting valve lever T should be vertical, which places valve S in forward position, opening both ports N and O.

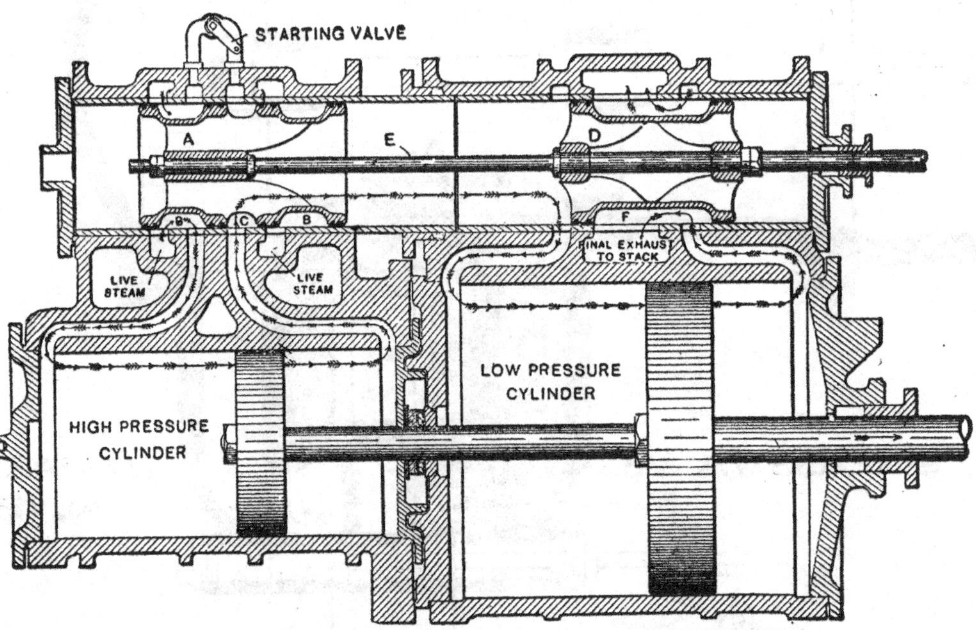

STEAM DISTRIBUTION IN TANDEM COMPOUND CYLINDERS

FIGURE 147

For Fig. 141 test, the starting valve S is in position as shown in Fig. 145, but having high-pressure valve A on center, by-pass valve M removed. For Fig. 143 test, valves A and S are in position as shown in Fig. 145, but having by-pass valve M replaced.

Starting Valve in Position for Working Compound. Fig. 146, same section as Fig. 145. Both by-pass valves in place. Lever T in back position, so starting valve S covers port O.

For Fig. 140 test, starting valve S as in Fig. 146. The high-pressure valve A on center.

For Fig. 143 test, valves A and S in position as shown in Fig. 146.

For Fig. 144 test, starting valve S as in Fig. 146. High-pressure valve A in forward motion.

The Baldwin Tandem Compound. In this type of locomotive, designed in 1902, principally for heavy freight service, four cylinders are used, with a high and low-pressure cylinder and cylindrical valve chest on each side. The high-pressure cylinder is placed in front of the low-pressure, both having the same axis; that is, the center of the low-pressure cylinder extended becomes also the center of the high-pressure.

Fig. 147 is a sectional elevation of the cylinders, valve chests and valves. The arrows show the distribution of the steam.

Each cylinder with its valve chest is cast separately and is separate from the saddle. The steam connections are made by a pipe from the saddle to the high-pressure valve chest, and the final exhaust takes place through an adjustable connection between the low-pressure cylinder and the saddle casting. The valve, which is double and hollow, admits steam to the high-pressure cylinder, and at the same time distributes the high-pressure exhaust from the front end of the high-pressure cylinder to the back end of the low-pressure cylinder or vice versa, as the case may be, without the necessity of crossed ports. As shown in the accompanying diagram, Fig. 147, A is the high-pressure valve by which steam is conducted from the live-steam openings through external cavities B and B to the high-pressure cylinder. The exhaust from the high-pressure cylinder passes through the opening C to the steam chest, which acts as a receiver; D is the low-pressure valve connected to the high-pressure valve by valve rod E. This valve in its operation is similar to the ordinary slide valve. The outside edges control the admission, and the exhaust takes place through the external cavity F. The starting valve connects the live-steam ports of the high-pressure cylinder

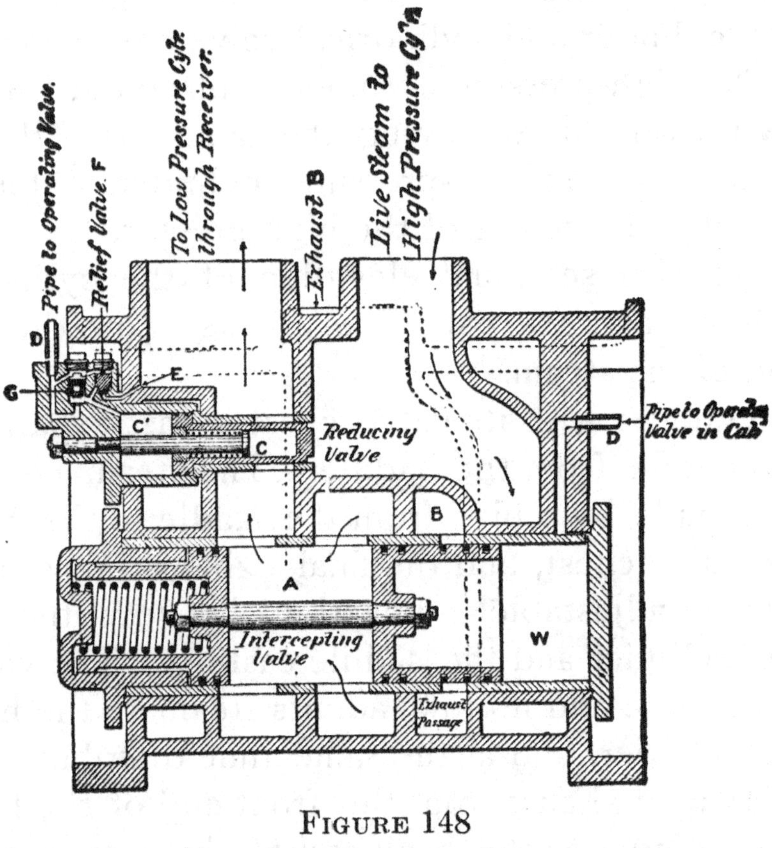

FIGURE 148

The Cross Compound. The cross compound locomotive has two cylinders, one on each side, with an intercepting valve, so arranged that the engineer can work the engine either simple or compound. When the engine is worked as a simple engine, the pressure of the steam that is admitted to the low-pressure cylinder is controlled by an automatic reducing valve in such a manner that it shall bear the same ratio to the pressure of steam admitted to the high-pressure cylinder as the volume of the high-pressure cylinder bears to the volume of the low-pressure cylinder. Unequal strains are thus avoided. As previously stated, a compound locomotive should never be worked as a simple engine, except in starting a heavy train, or when there is danger of getting "stuck" on a heavy up grade.

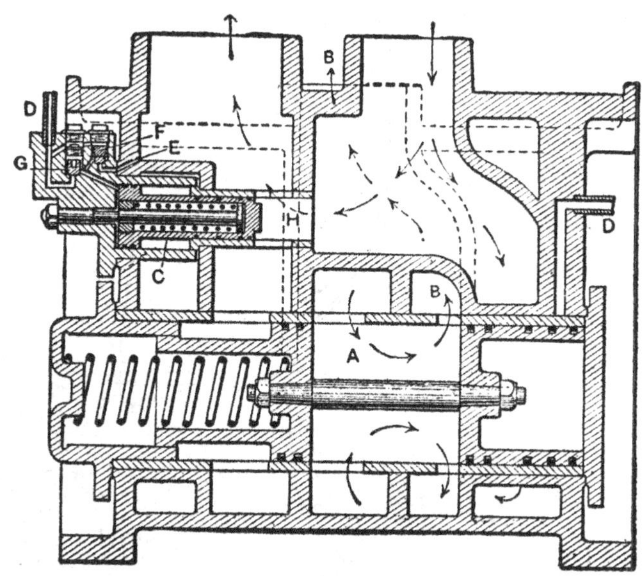

FIGURE 149

The Baldwin Two-Cylinder Compound. The essential features of this design, brought out in 1898, are the intercepting and the reducing mechanisms. These, when in normal position, permit the locomotive to operate by single expansion, and so continue until changed to compound. The engine is therefore readily started at any position of the crank.

In the diagrams, Figs. 148 and 149, A is a double piston intercepting valve, located in the saddle casting of the high-pressure cylinder. In one direction the movement is controlled by a spiral spring, in the other by steam pressure. The function of the intercepting valve is to cause the exhaust from the high-pressure cylinder to be diverted, at the option of the engineer, either to the open air when working single expansion, or to the receiver when working compound. C is a reducing valve, also placed in the saddle casting of the high-pressure cylinder, and like the intercepting valve is moved in one direction by a spiral spring, and in the opposite direction by steam pressure. The function of this valve is, in its normal position, to

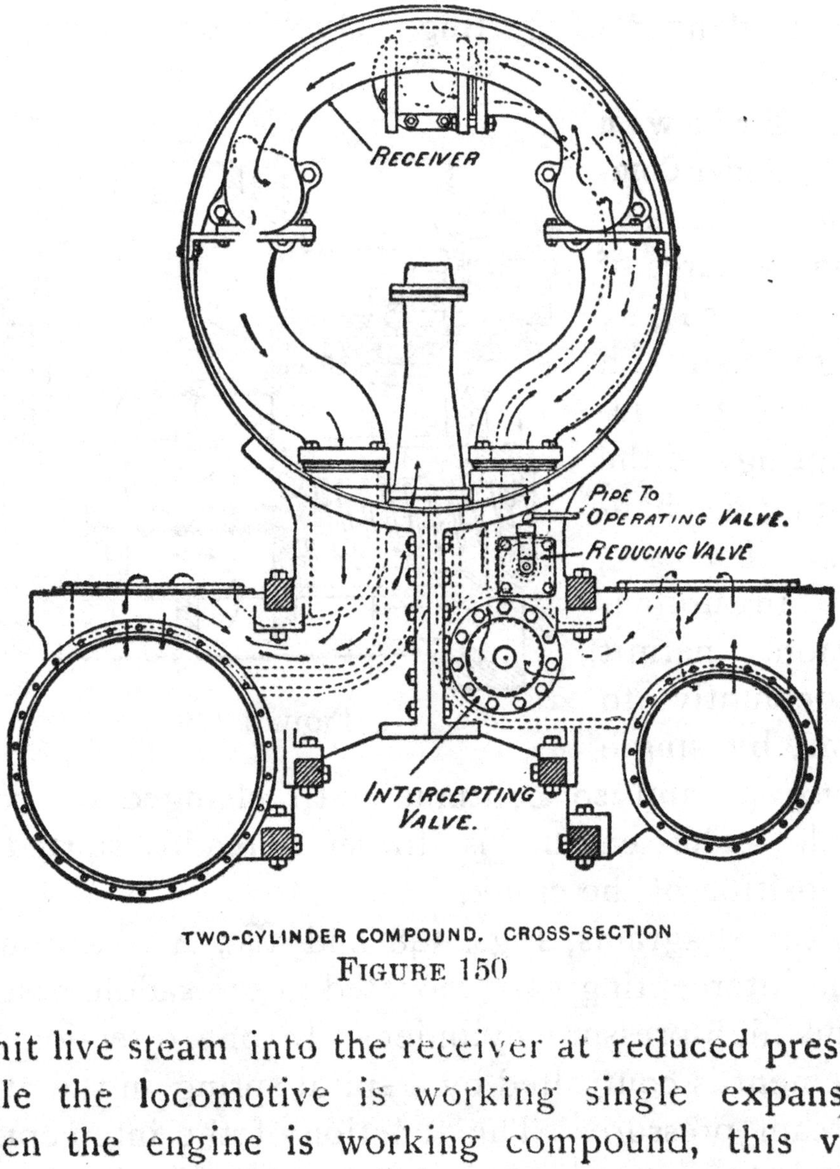

TWO-CYLINDER COMPOUND. CROSS-SECTION

FIGURE 150

admit live steam into the receiver at reduced pressure while the locomotive is working single expansion. When the engine is working compound, this valve automatically closes, as it is evident that there is no further need of live steam in the receiver.

A further function of the reducing valve is to regulate the pressure in the receiver so that the total pressure on the pistons of the high and low-pressure cylinders may be equalized.

The steam for controlling the operation of both intercepting and reducing valves is supplied through pipes D from the operating valves in the cab. When not permanently closed by pressure in the pipes D, the

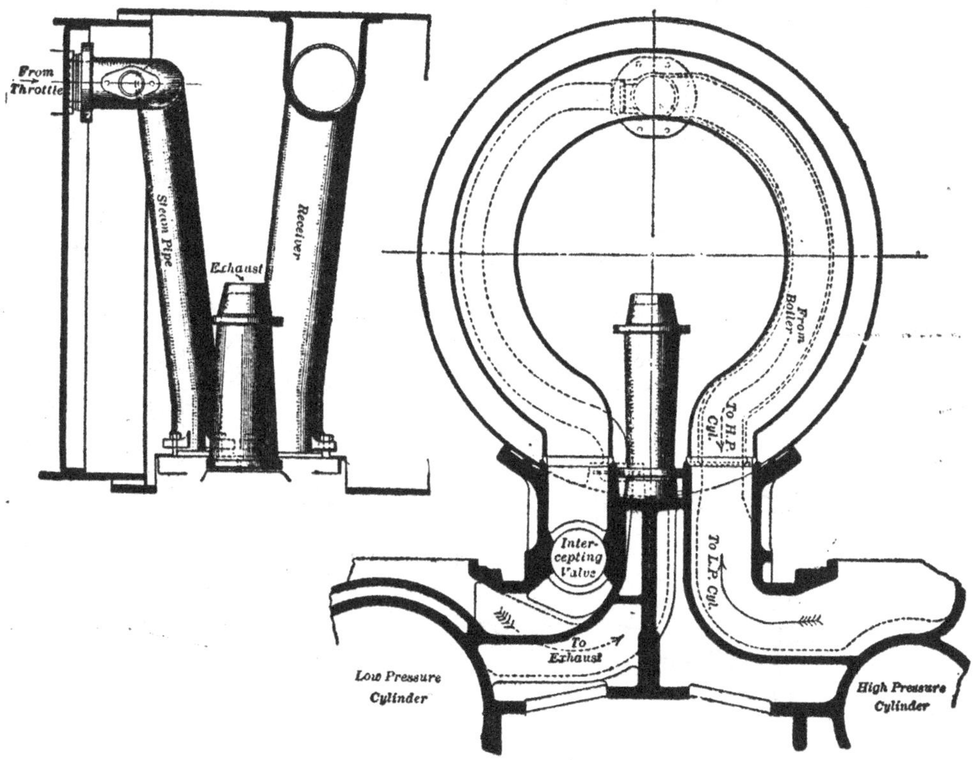

FIGURE 151

reducing valve C is operated automatically by the pressure in the receiver. To this end the port E is provided, communicating with the receiver, and the space in front of the reducing valve; as the pressure rises the steam acts on the large end of the reducing valve, causing it to move backward and close the passage H, through which steam enters the receiver, and thus prevent an excess pressure of steam in the low pressure cylinder.

Poppet valves F and G are placed in connection with port E, one to prevent the escape of steam from the receiver to pipe D when the locomotive is working single expansion, and the other to close the passage from pipe D to the receiver when working compound. Normally the lever of the operating valve in the cab is in the position marked "simple." In this position no steam is allowed to enter the pipes D, and no pressure

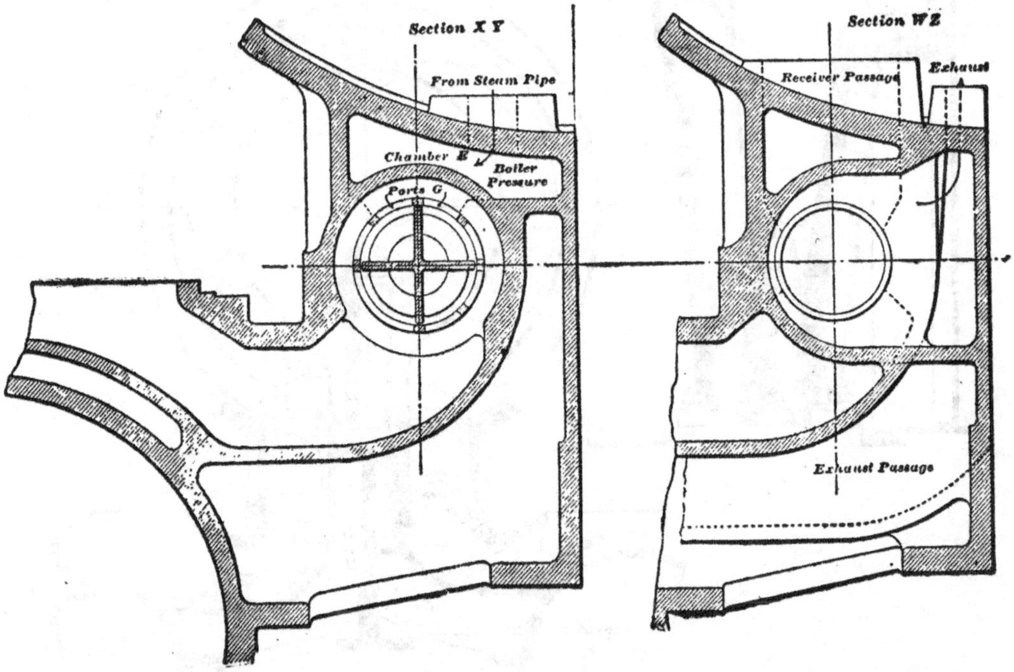

FIGURE 152

will be exerted on the intercepting and reducing valves in opposition to the springs, and they will assume the positions shown in Fig. 148.

The ports of the intercepting valve A stand open to receive the exhaust steam from the high pressure cylinder, and deliver it through the exhaust passage B to the atmosphere.

The reducing valve is open, admitting live steam through passage H to the receiver, and from thence to the low-pressure cylinder.

The receiver pressure is governed by the automatic action of the reducing valve, as previously explained. In this way the engine can be used single expansion in making up and starting trains, for switching and slow running.

At the will of the engineer the operating valve in the cab is moved to the position marked "compound." This admits steam to the pipes D, and through them to the valve chambers W and C^1, changing the intercepting and reducing valves instantly and noiselessly to the positions shown in Fig. 149. The exhaust from the high-pressure cylinder is diverted to the receiver, the admission of live steam to the receiver is stopped by the closing of the passage H, and the locomotive is in position to work compound.

Both valves are of the piston type, with packing rings to prevent leakage. This insures an easy movement of the valves, and prevents the hammering action common to valves of the poppet type when automatically operated.

Schenectady Cross Compound. The American Locomotive Company kindly furnish the following description of the two-cylinder cross compound engine as built at their Schenectady works.

Figure 151. A sectional view through the smoke arch and cylinder saddles, showing the steam passages, receiver, and the location of the intercepting valve in the low-pressure cylinder saddle.

Fig. 152. A transverse section through the low-pressure cylinder saddle XY and WZ. Section XY shows the passages for admitting live steam into the low pressure cylinder, and section WZ the outlet passage from separate exhaust valve to the exhaust pipe.

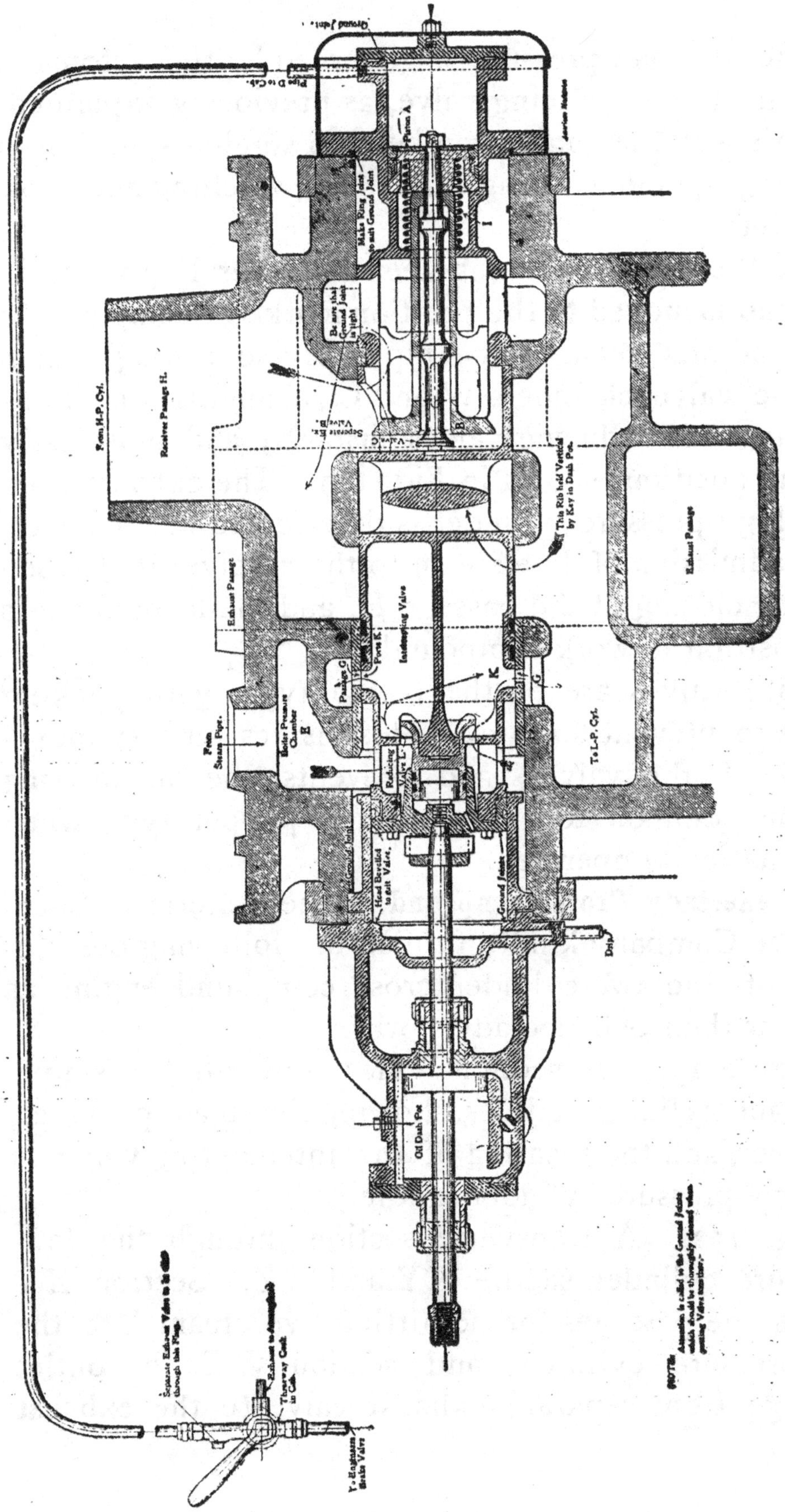

FIGURE 153

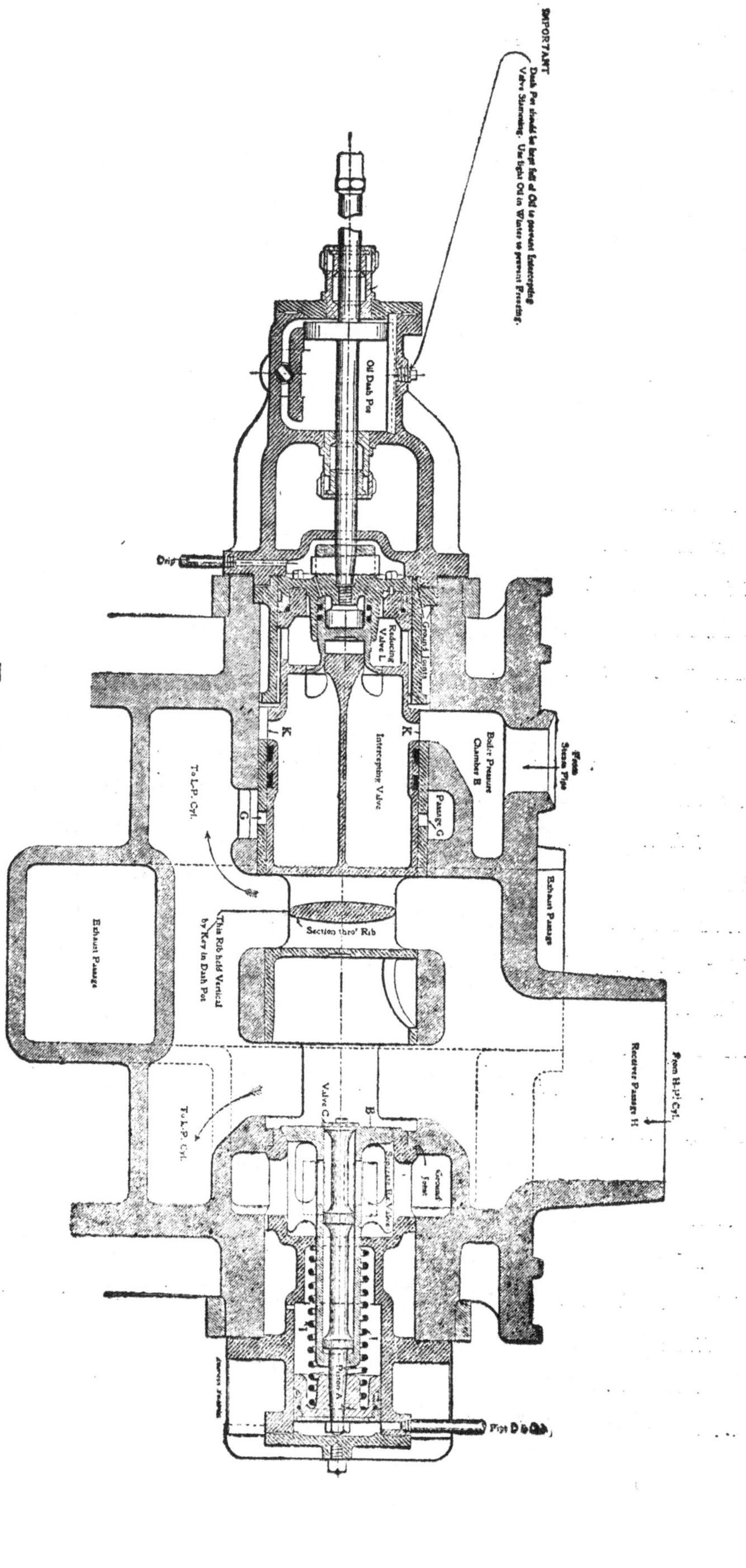
Oil Dash Pot
Reducing Valve L
Intercepting Valve
Boiler Pressure Chamber B
Exhaust Passage
To L.P. Cyl.
Section thro' Rib
This Rib held Vertical by Key in Dash Pot
Exhaust Passage
Receiver Passage H
To L.P. Cyl.

FIGURE 154

Fig. 153. A vertical section through the low-pressure cylinder saddle and intercepting valve, showing the intercepting and separate exhaust valves in the position taken when engine is working simple.

Fig. 154. The same section as Fig. 153, but shows the position of the intercepting and separate exhaust valves, when the engine is working compound. With the arrangement of valves shown in these figures the engine can be started and run either compound or simple, and can be changed from compound to simple, or from simple to compound, at the will of the engineer.

General Description. As the throttle is opened, steam from the boiler, through the dry pipe, is admitted directly to the high-pressure steam chest, and at the same time to chamber E, surrounding the reducing valve L, Figs. 153 and 154.

The exhaust from the high-pressure cylinder, by means of the receiver pipe, passes to chamber surrounding the intercepting valve, and thence to the low-pressure steam chest when working compound, intercepting valve in position shown in Fig. 154, or to the atmosphere, through separate exhaust valve and stack, when working simple, valve in position shown in Fig. 153.

The low-pressure exhaust passes directly to the stack at all times.

The intercepting valve opens and closes the connection between the two cylinders.

The separate exhaust valve opens and closes the connection between the high-pressure cylinder and the atmosphere.

The function of the reducing valve, which operates only when the engine is working simple, or starting.

SCHENECTADY 2-CYL. COMPOUND, BUILT FOR THE BUTTE, ANACONDA AND PACIFIC RAILWAY

PASSENGER LOCOMOTIVE, BUILT FOR CLEVELAND, CINCINNATI, CHICAGO AND ST. LOUIS RAILWAY

is to control the admission of steam from the boiler to the low-pressure cylinder, in order that the pressure of steam admitted to the low-pressure cylinder shall have the same ratio to the steam in the high-pressure cylinder as the volume of the high-pressure cylinder is to the volume of the low-pressure cylinder.

The oil dash pot insures a steady movement of the intercepting valve.

The intercepting and reducing valves operate automatically by means of the steam pressure acting on the difference of areas of the ends of the valves. The movement of the reducing valve is cushioned by the small air dash pots shown. The separate exhaust valve is operated by the engineer, by means of a three-way cock in the cab. To open the separate exhaust valve, the handle of the three-way cock is moved to the position provided for admitting pressure against the piston A, Fig. 153. Moving the handle in the opposite direction relieves the pressure against A, and the spring, which is shown in the figure, shuts the valve. The separate exhaust valve can be so connected as to operate either by air or steam.

Operation, Starting Simple. The handle of the three-way cock in the cab is moved by the engineer so as to admit pressure through the pipe D against the piston A, forcing it and the valves B and C to the position shown in Fig. 153. As the throttle is opened, steam is admitted directly from the boiler into the passage E, forcing the intercepting valve into the position shown (Fig. 153), thence the steam passes through the intercepting valve by the ports K K, and the passage G G, through the reducing valve to the low-pressure steam chest; at the same time steam from the boiler is admitted directly, by means of the steam pipe, to the

PASSENGER LOCOMOTIVE, BUILT FOR LAKE SHORE AND MICHIGAN SOUTHERN RAILWAY

Tandem Compound, Built for N. Y. C. and H. R. R. R.

high-pressure steam chest. The exhaust from the high-pressure cylinder passes to the atmosphere by means of the receiver passage H and the separate exhaust valve B. Steam from the low-pressure cylinder is exhausted directly to the atmosphere.

To Change from Simple to Compound. Having started simple, to change to compound, the handle of the three-way cock in cab is turned so that pressure is released from the piston A. The separate exhaust valve will then be closed by the spring I. The pressure in the receiver, due to the exhaust from the high-pressure cylinder, will rise and force the intercepting valve to the left, that is, to the position shown in Fig. 154, thereby opening the passage for the exhaust steam, from the high-pressure cylinder, through the receiver, to low-pressure steam chest. The movement of the intercepting valve to the left also closes the passage G G, thereby shutting off the admission of steam directly from the boiler to the low-pressure steam chest.

Starting Compound. To start the engine compound the separate exhaust valve is left closed as in Fig. 154. As the throttle is opened the steam pressure in the passage E will force the intercepting valve to the right or to the closed position; at the same time steam directly from the boiler will be admitted to low-pressure steam chest through ports K K and passage G G. The high-pressure cylinder will exhaust into the receiver until the pressure is sufficient to force the intercepting valve to the left, as shown in Fig. 154, when the engine will work compound. The change to compound working takes place at from one-half to three-quarters of a revolution of the driving wheels.

Compound to Simple. With the engine working compound, if the engineer wishes to run the engine simple to prevent stalling on a heavy grade, the handle of the three-way cock should be placed in same position as for starting simple. This opens first the small bleeding valve C, Figs. 153 and 154, and then the separate exhaust valve. The bleeding valve relieves the pressure and thus permits the main valve B to be operated more easily. As soon as the separate exhaust valve is open, the pressure in the receiver drops and the intercepting valve is forced against the seat to the right, by means of the pressure in chamber E, and the engine works simple as before. Engines should be worked simple no longer than absolutely necessary.

Lubrication. A pipe from the sight feed lubricator located in the cab leading directly to chamber E is provided, by means of which both the intercepting and reducing valves are lubricated. One drop per minute is sufficient for these parts. A small oil cock in three-way cock, located in cab, provides for lubricating the separate exhaust valve and attendant parts, and oiling once a day with a small quantity of cylinder oil provides sufficient lubrication.

When using steam it is good practice to feed about two-thirds of allowance of cylinder lubrication to H.-P. cylinder. When drifting down long grades this should be reversed, on account of the larger surface to be lubricated on L.-P. side. Always run with lubricator steam valve wide open.

By-Pass Valves. Some of the compound locomotives recently built are equipped with by-pass valves, provided to admit of engines drifting more freely. These valves, more particularly on the low-pressure side, should be examined occasionally, by removing the cap, to insure that they are in good working order.

FOUR CYLINDER ARTICULATED COMPOUND, BUILT FOR BALTIMORE AND OHIO R. R.
HEAVIEST AND MOST POWERFUL LOCOMOTIVE EVER BUILT

On new engines the by-pass valves should be cleaned frequently, as their free movement is liable to be hindered by gumming or the presence of core sand.

Should a by-pass valve become broken or in any way defective, take off the valve body and insert a blind gasket between it and the cylinder.

Carrying Water. Most of the later compound locomotives are equipped with piston valves, and it is very necessary that the cylinders should be kept free from water. Great care should be taken to open cylinder cocks when starting and before opening throttle after drifting down grade. Careful attention should also be given to avoid carrying water too high in boiler. Carrying water high in the boiler, and thus causing wet steam in cylinders, is injurious to compound locomotives, no matter whether slide valves or piston valves are used.

Oil Dash Pot. This should be kept full of oil, to prevent intercepting valve from slamming. Breakages of intercepting valves are nearly always due to neglect of this rule.

Dash pots should be filled with common car or engine oil, thinned with kerosene when necessary, in winter.

The dash pot stuffing boxes should be kept packed, to avoid leakage of oil.

Drifting. In drifting, the three-way cock should be in simple position whenever it can be done without too much loss of air by leakage of separate exhaust valve or piping. Most of the recent compound locomotives are provided with a small drifting valve, in main throttle valve, so arranged that it can be opened with a slight movement of the throttle lever. It is considered good practice to admit a little steam to cylinders when drifting, through this valve, or, if not provided

TANDEM COMPOUND, BUILT FOR CAPE GOVERNMENT RAILWAYS, SOUTH AFRICA

with a small drifting valve, by a slight opening of main throttle.

Examination. Enginemen should ascertain if separate exhaust valve is in good working condition before starting out with train, by trying the engine simple and compound before coupling to the train. The separate exhaust valve should be examined at intervals, so that the spring and other parts are kept in proper condition. Should the engine refuse to move after the throttle is opened, it will usually be found that it stands on center on high-pressure side (in position to take steam on low pressure side), and it will be due to either the intercepting or reducing valve sticking, which is always the result of lack of lubrication for intercepting valve, or carrying too much water in the boiler. Which of these valves are sticking can be ascertained from the position of the intercepting valve stem. In starting the engine, if the intercepting valve stem extends clear out about 7 in., it would be the intercepting valve, and unless some of the ports are broken a slight tap on the end of the stem, with throttle open, would send it ahead. If it was found that the stem had already moved ahead so that it extended out about 3 in., it would be the reducing valve. Usually one or two sharp blows on the intercepting valve back head, with throttle open, will loosen it. In either case live steam would then be admitted to low-pressure cylinder for starting.

Should the engine refuse to work compound after the three-way cock had been placed in compound position, and continue to work as a simple engine, it would indicate that the separate exhaust had not closed. This trouble can usually be traced to enginemen using engine oil for lubricating separate exhaust valve chamber, and can sometimes be overcome by a dose of kerosene, which should in all cases be followed up with valve oil.

Relief Valves. Combined pressure and vacuum relief valves on low-pressure steam chest and single-pressure relief valves on low-pressure cylinder heads should be set at 45 per cent of the boiler pressure, and the high-pressure cylinder head relief valves set at 20 lbs. above boiler pressure.

Dampers. Dampers should be closed when drifting down long grades.

QUESTIONS

381. What is the principal object in the compounding of locomotives?

382. Name two sources of economy in compound engines.

383. Why is there a constant loss of heat in the single cylinder engine?

384. How is the expansion of the steam divided in the compound locomotive?

385. How should the cylinders of a compound engine be proportioned regarding size?

386. What other problems are before the designers of compound locomotives?

387. How many types of compound locomotives are in use in this country?

388. Describe briefly the Vauclain compound.

389. What kind of an engine is the balanced compound?

390. How are the cylinders of the tandem compound located?

391. How many cylinders has the cross compound?

392. What kind of valve gear is used on compound locomotives?

393. What were some of the objects aimed at in designing the Vauclain compound?

394. How many and what type of valves are used on the Vauclain compound?

395. What kind of packing rings are used on this valve?

396. When is the Vauclain valve motion direct acting?

397. When is it indirect?

398. In setting these valves, what ports are to be considered?

399. Of what material are the pistons made?

400. In starting these engines with full trains, what is necessary?

401. How is this accomplished?

402. What is the starting valve, and what is its function?

403. How is it operated?

404. What rule should be observed regarding this valve?

405. What provision is made for taking care of water that finds its way into the cylinders?

406. What is the first thing an engineer should learn, in the operation of a compound locomotive?

407. How is the quadrant of the Vauclain compound made, with reference to point of cut-off?

408. What rules should be observed when starting the Vauclain compound?

409. When should the reverse lever not be hooked up?

410. Should the starting device be used when the train is in motion?

411. When is it allowable to use the starting device while the train is in motion?

412. How should the fire be carried in the Vauclain compound?

413. Where is the most economical point of cut-off for a single expansion engine?

414. Where is the most economical point of cut-off for a compound locomotive?

415. What should be done when starting the Vauclain compound?

416. What should be done with the reverse lever as the speed of the engine increases?

417. What should be done on a slightly descending grade?

418. What should be the position of the starting valve lever, when throttle is closed?

419. If there is danger of stalling on a heavy upgrade what should be done?

420. What is one of the legitimate advantages of the compound locomotive?

421. What advantage has the boiler of a compound locomotive over the boiler of a simple engine?

422. What is the ideal type of engine, whether stationary or locomotive?

423. How may this ideal be reached?

424. What has always been a serious problem for locomotive builders?

425. How are the cylinders of the Baldwin balanced compound located?

426. What type of valve is used on these engines?

427. Where are the valves located?

428. Where are the high-pressure cylinders located?

429. At what angle are the cranks set?

430. Describe briefly the action of the steam in this engine.

COMPOUND LOCOMOTIVES

431. How are the cylinders of the American Locomotive Company's balanced compound located?

432. How is a uniform turning moment attained in this engine?

433. Mention the advantages that the balanced compound possesses over other types of compound locomotives.

434. Why does the tandem compound appear to be the ideal locomotive?

435. What is one of the main objections to this type of compound locomotive?

436. What kind of handling does a compound engine require?

437. What knowledge is necessary for the engineer in order that he may successfully operate a compound engine?

438. What can be said regarding the tandem compound built by the American Locomotive Co.?

439. How are the valves arranged on this engine?

440. What is the function of the starting valve?

441. How should a compound locomotive be lubricated?

442. How are the cylinders placed in the Baldwin tandem compound?

443. What about the cylinders and valve chests of this engine?

444. What kind of a valve has this engine?

445. How many cylinders has a cross compound, and how are they located?

446. What is the purpose of the intercepting valve?

447. What is the function of the automatic reducing valve?

448. How is the steam for operating these valves supplied?

449. How is the receiver pressure governed?

450. What are the by-pass valves for?

451. What precautions should be observed regarding water on these engines?

452. What about the oil dash pot?

453. What should be done when drifting?

454. What should be done with the separate exhaust valve?

455. Should the engine refuse to move when the throttle is opened, what would be the probable cause?

456. How may it be ascertained which one of these valves is stuck?

457. How may the stuck valve be loosened?

458. In what position should the dampers be when drifting?

OTHER BOOKS FROM CGR PUBLISHING AT CGRPUBLISHING.COM

Ultra Massive Video Game Console Guide Volume 1

When Trolleys Ruled the Earth: A Photographic History of Streetcars...

Complete Works of Edgar Allan Poe Volumes 1 - 10

Classic Cars and Automobile Engineering: Volumes 1-5

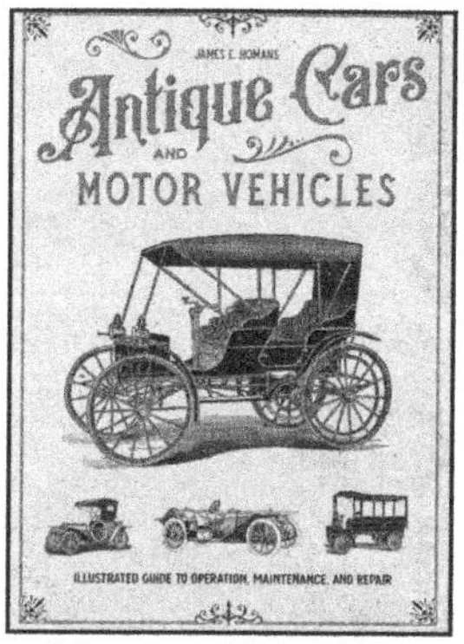

Antique Cars and Motor Vehicles: Illustrated Guide to Operation...

Chicago's White City Cookbook

The Clock Book: A Detailed Illustrated Collection of Classic Clocks

The Complete Book of Birds: Illustrated Enlarged Special Edition

1901 Buffalo World's Fair: The Pan-American Exposition in Photographs

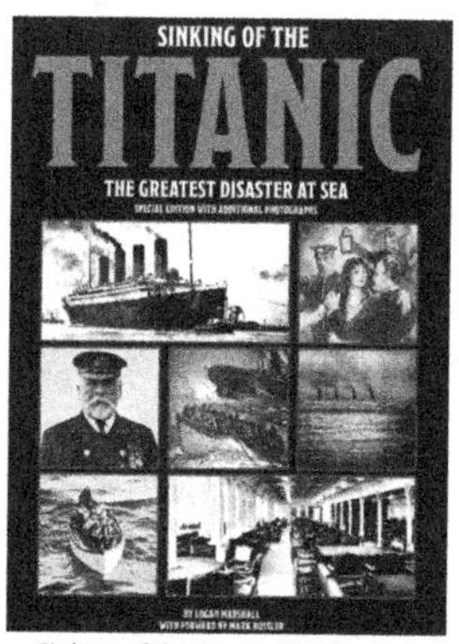

Sinking of the Titanic: The Greatest Disaster at Sea

Gustave Doré's London: A Pilgrimage: Retro Restored Special Edition

Milton's Paradise Lost: Gustave Doré Retro Restored Edition

The Art of World War 1

The Kaiser's Memoirs: Illustrated Enlarged Special Edition

History in the Age of Vikings Series Volumes 1-3

The Complete Butterfly Book: Enlarged Illustrated Special Edition

\- MAILING LIST -

JOIN FOR EXCLUSIVE OFFERS

www.CGRpublishing.com/subscribe

Made in the USA
Monee, IL
19 November 2025

35557157R00190